Economic Evaluation
of Oil & Gas Properties

# ECONOMIC EVALUATION OF OIL & GAS PROPERTIES

Richard J. Miller

PB
PENNWELL
BOOKS

**Disclaimer**

The recommendations, advice, descriptions, and methods in this book are presented solely for educational purposes. The author and publisher assume no liability whatsoever for any loss or damage that results from the use of any of the material in this book. Use of the material in this book is solely at the risk of the user.


PennWell Books, LLC
10050 E 52nd Street
Tulsa, Oklahoma 74146 USA

866.777.1814
sales@pennwellbooks.com
www.pennwellbooks.com

Publisher: Matthew Dresher

Library of Congress Cataloging-in-Publication Data

Names Miller, Richard J. (Consulting engineer) author

Title Economic evaluation of oil & gas properties [Richard Miller].

Tulsa, Oklahoma. PennWell Books, 2025. Includes bibliographical references and index.

Identifiers LCCN 2025000556 (print) LCCN 2025000557 (ebook) ISBN 9781955578301 hardcover ISBN 9781955578318 epub

Subjects LCSH Oil fields--Valuation--United States Gas fields--Valuation--United States

Classification LCC TN871 .M485 2025 (print) LCC TN871 (ebook)
DDC 622.3382--dc23eng20250320

LC record available at https://www.lccn.loc.gov2025000556

LC ebook record available at https://www.lccn.loc.gov2025000557



Printed in the United States of America

1 2 3 4 5 29 28 27 26 25

# Contents

## Chapter 5



## Chapter 6



## Chapter 7



## Chapter 8



## Chapter 9

## Chapter 19



## Chapter 20



## Appendices

# Acknowledgements

The content of the book is derived from over 50 years of property evaluation practice, publication of book sections, professional papers and studies, and teaching university classes and industry seminars. Accomplishing the above is the result of education and experience for which many friends, colleagues, and students share responsibility, but I would like to acknowledge a certain few: Langdon B. Taylor, Professor of Petroleum Engineering, New Mexico Institute of Mining and Technology, who taught me petroleum engineering and a certain amount of humility; Jim Smith, my first boss, from whom I learned the Jim Smith Rule; Nico van Wingen, Professor of Petroleum Engineering, University of Southern California, whose lectures provided the economic context for all of that production and reservoir engineering education; Milton Lewis, who gave me a start in consulting practice; Arthur Spaulding, Chairman, Western States Petroleum Association, who provided the opportunity to examine discount rates and fair market value in depth; Clayton Clement, Attorney in Santa Rosa, CA, who taught me the legal applications of evaluation work; and Robert Wayne, Attorney in Williamsport, PA, who introduced me to the Marcellus Shale, hunting clubs, and the "perch." I also would like to acknowledge my good friend, respected petroleum engineer and Texas rancher Dan Olds, who read through every page and offered many useful comments and suggestions.

However, this book is dedicated to two people who made it all possible.

Catherine McGregor Chisholm Walker McAlpine, my Scottish grandmother, who made sure that I was pointed in the right direction and who checked up now and then to be sure I stayed that way.

And, most especially, my wife of 40+ years, Carol, who has been continually supportive in many more ways than I can count.

# Introduction

Any discussion of the evaluation of oil & gas properties is necessarily complex. Even a discussion of the sub-topic of economic evaluation can become complicated. So, it may be useful to start with a personal anecdote. I went to work for Texaco in California in June 1967. I had my brand-new degree in hand and was sure that I knew all there was to know about drilling for and producing oil & gas, something which I was intent on proving at the earliest opportunity. After working in a small field office for about six months, I assembled the paperwork to support a project to restore numerous idle wells to production, including a description of the required remedial work, new equipment, estimated costs, expected work schedule, anticipated production, and the potential income stream. The District Engineer at the time was a fellow named Jim Smith. While I explained the project, he listened attentively, asked a few questions, and then pointed out that the planned investment to do the work was substantial but the economic return (profits, length of investment recovery time, etc.) was not very good. I agreed but hastened to note that the oil production would be considerably higher than the current production and the potential future oil recovery would increase substantially. After all, that was the purpose of all that education. It was then patiently explained to me that:

"*We are not here to make oil, we are here to make money.*"

This is known as the **Jim Smith Rule** and has been the guiding principle of my work for 50+ years.

This simple premise is the reason why an economic evaluation is necessary when developing oil & gas properties. However, as with so many simple premises, implementation requires elaboration. Every decision made regarding the development and maintenance of production must consider the economic costs and potential benefits of the decision. This is true regardless of whether the project involves drilling a new well, changing a pumping unit, consideration of a stimulation method, installing an Enhanced Oil Recovery (EOR) project, or buying a property. The expected benefits may include:

- increased production rate,
- expanded access to production, or
- reduced operating costs.

These anticipated benefits may be offset by a need for new capital investment or the acceptance of increased operating costs in exchange for an increase in production, but the result must be optimal economic return.

# Definition of Property

Before going further, an understanding of property may be useful. As used in this discussion, a property is any discrete source of production and income. It can be a "lease" or "fee" property with several wells, a group of leases or part of a field, or it can be a single well [see Real Estate, etc., p. 11]. On the other hand, it can also be a project or a combination of properties owned by a company or individual and operated as an entity, such as a unit organized for EOR purposes. A property can be as unique as a royalty interest in three wells in Lycoming County, Pennsylvania, or the property could also be a "project" such as a proposed or functioning production source composed of multiple properties or (as in the THUMS Unit or Elk Hills field, both in California) separate fault blocks or stratigraphic reservoirs. The governing criteria are that the property have a description and the source of the production and income be identifiable, definable, and quantifiable. *In this discussion, the terms "property" and "project" are used interchangeably as the objects of evaluation.*

# 1

# PURPOSES OF ECONOMIC EVALUATION

The objective of petroleum property evaluation is to examine, define, and implement methods and procedures for developing and extracting crude oil and natural gas as well as any associated products to optimize profits and, at the same time, to obtain an economic return that is commensurate with the risk incurred in making the investment. *The term "optimize" is used rather than "maximize" because there are factors other than financial considerations that can influence both the decision to make an investment in an oil and gas project and the return to be obtained from the investment; the project may not provide the maximum economic return, but may provide an optimal return when other considerations are included.*

Hereafter, to avoid repetition of the phrase "oil and/or gas," the terms **oil, hydrocarbon,** or **petroleum** are used as a reference to both oil & gas, unless specific reference to one or the other is necessary.

The economic evaluation of a property or project consists of two broad functions:

- Estimation of the potential volumes of producible oil and/or gas attributable to a property or project, and derivation of one or more schedules of recovery of the producible volume.
- Estimation of the economic benefit of the anticipated production.

These are not separate functions. Some aspects of the first are necessary to accomplish the second objective and, in that respect, the two functions meld at the boundary so that one influences the other. The first part of this discussion will deal **briefly** with estimating the amount of producible oil & gas along with estimating the future production schedule. The second part will define the procedures for estimating the economic benefit of future production.

There are many reasons for estimating the economic value of an oil & gas project. A number of methods can be used to make the estimate. However diverse the reasons for an evaluation, they can generally be placed in one of two groups: Investment Analysis or Required Valuation.

# Investment Analysis

For the most part, **Investment Analysis** is done for internal business purposes, such as planning of capital expenditures, or for external applications such as financing support and sale or acquisition.

## Planning for Capital Expenditures and Budgeting

Investment analysis for capital expenditures and budgeting could be anything from the evaluation of the potential economic benefit from a workover or equipment upgrade to drilling a new well or acquiring a producing property. Regardless of the purpose, the practical aspects are identical. The evaluator is attempting to determine the additional benefit in terms of future income, profit, and return-on-investment, etc., that could be obtained by making a specific expenditure.

> As an example, if it will cost $100,000 to do a well workover, acidize a well, or open a new production interval, will the additional production result in an increase in revenues sufficient to pay back the $100,000 in a reasonable time and provide a return on the investment that compensates for the financial risk incurred by making the investment? Further, is the return on this investment sufficient to justify the use of funds for this project rather than another similar project?

If financial recovery of the investment requires three years, and there is a measurable possibility that the workover may not increase production, then there is some risk associated with waiting three years to recover the investment and the likelihood of a less-than-expected income stream. Economic evaluation attempts to determine if the future return will be enough to compensate for the combined risks of the investment, and whether the investment should be made. The same principles apply regardless of the amount of the expenditure.

## Comparison of Investment Opportunities

The relation between benefits and costs can be measured in various ways. Among these are undiscounted net profit, net present value (NPV), internal rate-of-return (ROR), and various other methods that will be discussed in detail in later sections. Each of these methods provides a different perspective on investment analysis. Each method affords the evaluator a means to measure the relationship between expected future income and the capital investment required for the project, and to make rational comparisons when selecting projects for completion.

As an example, if a new well is proposed which could result in a net profit of $1 million and a ROR of 19%, this project might be compared to other projects and also to company criteria for net profit and ROR. If the company requires a minimum ROR of 15%, the project might be accepted. If, on the other hand, the net profit was $1 million but the ROR was only 9%, the project ROR would not only fail to meet company guidelines, but might be less than could be obtained on investments without a comparable risk.

## External Applications (Financing, Acquisitions)

In the past, commercial banks, trust companies, and insurance companies that extended debt financing, as well as equity underwriters and investors, generally required that borrowers provide economic evaluations of the properties offered as collateral, or at least provide current economic evaluations as a measure of income potential. These evaluations were used to establish both the collateral value of the properties and the ability of the properties to repay the debt on these *production loans*. Energy lending practices have changed over time, but it is still common for lenders to require timely evaluations of a borrower's properties. It is also not uncommon for a lender to prescribe certain criteria, usually product price projections, to be used in the evaluations.

Other, common reasons for an evaluation are to provide the seller of a property with an estimate of the value that could be received from a market sale, or provide a potential buyer with an estimate of how much the property is worth. These studies can be done internally by the buyer or seller but may be conducted by one or more third-party evaluators to estimate the value of a property, often under a variety of postulated economic and operational conditions.

# Required Valuations

These evaluations are done to estimate value for property tax, estate tax, gift tax, transfer tax, or eminent domain, among others. They are focused primarily on the "value" of the property for a specific purpose rather than the economic return that could be obtained. They are usually subject to an interpretation of value often referred to as "market value." Further, some aspects of the evaluation may be dictated by the controlling federal, state, or local jurisdiction. A special type of evaluation may be done for reporting to agencies, such as the Securities and Exchange Commission (SEC), where some of the parameters are defined by the agency. SEC evaluations are used for comparison among publicly traded companies and have only a tangential relation to economic value *per se*.

# 2

# SOURCE OF ECONOMIC BENEFIT IN OIL & GAS PROPERTIES

The source of value in oil & gas properties is the "right" to access, produce, and sell crude oil, natural gas and other products obtained from the property. This is a subtle concept that is often misunderstood by evaluators who refer to the acquisition or ownership of some volume of future production. It is also the genesis of the Jim Smith Rule. Oil & gas in the ground have no value, intrinsic or otherwise, unless the oil can be produced and sold at a profit. Property value derives from a future stream of income to be obtained from the production and sale of petroleum products. This stream of income has value to those willing to invest to obtain that income. A mineral right that can be valued can also be bought or sold, used as collateral for financing, or considered as an asset contributing to net worth.

Petroleum properties have three characteristics that complicate property valuation. First, the economic benefit of the property depends on production of commodities that are subject to all the vagaries of supply and demand for scarce resources, with the added complication that these commodities are often difficult to find, access, develop, produce, and market. Second, oil production is a long-term proposition; the recovery of expected oil and/gas volumes usually requires many years and often decades. Third, unlike surface land or developed real estate, which may be altered but not necessarily destroyed, the value of an oil property is based on an asset that is being irreplaceably depleted over time and which leaves no residual value.

It remains only to devise a means to estimate the size, shape and duration of the "income stream" and devise an appropriate method for determining the "economic benefit" of owning that income stream. The development of methods to estimate:

1. the volumes of oil & gas in the ground,
2. the portion of that volume that could be produced,
3. the rate at which that volume could be produced, and
4. the economic conditions that will apply to that property in the future allow an evaluator to estimate the income potential and value of the property.

## Concept of the Income Stream

The purpose for constructing an income stream is to attempt to model the economic benefit that may be anticipated from an individual oil property. The size, shape, and duration of the production and income stream are directly dependent on the physical and operational characteristics of the property, the economic conditions which currently exist and those which are expected to apply to the property in the future, and expenditures which might be expected to occur in order to take advantage of property characteristics and/or to respond to economic conditions. The construction of the income stream can be relatively simple or complicated depending upon the anticipated conditions.

## Concept of Ownership

**Oil & gas rights** – often referred to as "mineral rights" – are real estate. The term oil & gas rights is descriptive in that "ownership" is limited to the right to access and obtain, by some method, the hydrocarbons that may exist, but there is no ownership of the oil- and gas-in-place as such. The concept of the ownership of "mineral rights," as opposed to the ownership of the actual minerals, is important. Because of the difficulty of defining the volume and character of oil & gas deposits, and because the specific source of the hydrocarbons cannot be readily identified, oil & gas ownership refers only to the right to produce hydrocarbons underlying a defined surface location, not to ownership of the oil- and gas-in-place under that location.

There can be misunderstanding on this point.[1] The holder of a leasehold interest has only a "right-to-produce" subject to the terms of the lease – there is no ownership of the minerals. A disagreement may arise regarding a fee interest. The owner of a fee interest (see below, ownership of surface and minerals, p. 11) may consider that he "owns" the actual minerals. While in the case of a very large property, this may be true in a practical sense, it is not strictly true. All oil & gas properties are subject to a legal principle known as the **Rule of Capture**. As noted by Martin and Kramer, *The rule has been phrased as follows: "The owner of a tract of land acquires title to the oil and gas which he produces from wells drilled thereon, though it may be proved that part of such oil or gas migrated from adjoining lands."* cited from *Hardwicke, "The Rule of Capture and its Implications as Applied to Oil and Gas,"* 13 Tex. L. Rev. 391, 393 (1935).[2] It is true that issues arising from production across property lines are not common as properties become more consolidated, however, the rule and the related law still exist.

Mineral rights can be bought and sold just like surface land. They can also be leased to a third party, thereby creating a "leasehold," that the third party owns for some period described in the lease agreement, but the lease does not create an ownership interest. Leasehold interests in oil & gas rights can also be bought and sold. However, a purchase or sale of a leasehold interest does not transfer ownership of the minerals or the mineral rights – only the leasehold right to produce and sell the minerals is transferred.

## Concept of Economic Limit/Life

The **Economic Limit** is the production rate at which the net revenue from sale of the production of a volume of oil & gas equals the cost of production of that same volume. Where the purpose of owning an income producing oil property is to receive the net cash flow (income stream) from that property, then only the projection of future production and cash flow to the economic limit (the economic life) has tangible benefit. Continued production below the economic limit would result in a loss of income for every barrel or MCF produced and would reduce rather than add to value.

The Economic Limit can be difficult to calculate, particularly if operating costs are complex and/or closely related to the volume of production. The economic limit is typically expressed in terms of the primary product, whether oil or gas, but the calculation should include the revenue from all products and all the costs associated with production and sale of these products.

## Concept of Effective Date

Evaluation of oil & gas properties almost always requires the definition of an **effective date.** The effective date is the point in time that denotes the beginning of production activities and cash flows that follow. Any cash flows that may have occurred prior to the effective date are not reflected in the cash flows developed as part of the evaluation following the effective date. Internal evaluations may be done as of the beginning of the fiscal year or quarter for financial reporting, capital budgeting, or reports to lending institutions. They may also be done as of the beginning of a month for use in determining the desirability of an investment or for property purchase or sale. Valuations done for external purposes, such as ad valorem tax (usually January 1st) or estate tax (usually date of death), are done as of the date specified by the regulating party(ies).

## Concept of Present Value

An **income stream** is a series of payments to be received over time. **Present value** is another relatively simple concept that uses a **present value factor (PVF)** to *reduce an amount of income to be received in the future* (the schedule is irrelevant) to a comparable value today. Present value is the inverse of compound interest. Essentially the PVF is a measure of a person's perception of the value of a bird in the hand compared to a bird (or birds) in the bush. If offered X dollars 5 years in the future, possibly to be paid out as X/5 dollars each year, or Y dollars (usually a lesser amount) today, would the person rather have Y dollars today or a larger amount paid over 5 years? The present value of the deferred income to be received is reduced by an amount that varies with the evaluator's estimate of **opportunity cost**, the amount that he could have earned on an alternative investment of equal risk employing the deferred funds. In order to equate future income to the value today, a present-value factor is applied to the income stream.

# Oil & Gas Properties That Can Be Evaluated

The properties that may be subject to evaluation can be sorted into four broad categories:

## Producing

In this case the property (or a pooled unit that includes the property) has been developed through drilling and completion of wells and is producing hydrocarbons. There are sales of oil and/or gas, and the working interest owner is receiving income. The property may not necessarily be fully developed. It could be in an early stage of development and production or it could be a property that has been in continuous production for decades. It could be a small lease in north-central Pennsylvania or the San Andreas Unit in West Texas. Producing properties account for the large majority of all evaluations and probably 100% of the properties that are evaluated for internal purposes done for working interest owners.

There are, however, other forms of oil & gas property that can be evaluated depending on the purpose of the evaluation and the status of the interest owner.

## Leased and Partially Developed

The property may or may not be included in a pooled unit. There may be drilling permits which have been requested and/or approved, there may be a development plan in place, and there may be one or more wells drilled either on the property, or on a property with which it is pooled, but there may or may not be any production.

Evaluations for this category of property have become common in recent years, particularly for royalty owners, due to development of shale formations using horizontal drilling and hydraulic fracturing. The "shale boom" has led to leasing, drilling, and production in regions that had not previously been subject to oil & gas production in Pennsylvania, West Virginia, Ohio, North Dakota, and the Eagle

Ford region in south Texas, among others. The impetus for evaluations of these properties has largely come from Lessors who have found a need to consider, for the first time, the value of their oil & gas rights.

## Leased but Undeveloped

The oil & gas rights are leased. Geophysical surveys may have been conducted on the property. There may or may not be any well drilling permit requests submitted or approved for the property or nearby properties. The property may or may not be in a pooled unit. This is a subset of the leased and partially developed category. Valuations of this type are primarily for Lessors for financial planning and for transfer and estate tax purposes.

## Unleased Oil & Gas Rights

The value of unleased and undeveloped oil & gas rights depends on the purpose of the evaluation and the circumstances of the property. Such an evaluation is usually requested by the owner of the rights for financial or tax planning purposes. However, the value of undeveloped oil & gas rights can be required or requested for regulatory purposes such as eminent domain.

> As an example, a few years ago the State of California Department of Transportation requested an appraisal, for eminent domain purposes, of the mineral rights underlying a parcel of land which turned out to be on the edge of downtown Fresno. A look at historical drilling maps and other sources showed that there were no wells in the area and there had never been any production. Absent a basis for an appraised value, a nominal value of $10 per acre of oil & gas rights was assigned.

# 3

# REAL ESTATE, MINERAL RIGHTS, AND OWNERSHIP INTERESTS

Evaluation of oil & gas properties assumes a knowledge and understanding of the nature of the property and the extent of the property. It also requires a knowledge of the form of ownership of the property. Underlying the concept of property is the recognition of land as the primary source of value where a common understanding is that:

- Each parcel of land is unique in its location and composition
- Land is physically immobile
- Land is durable
- The supply of land is finite
- Land is useful to people

While historical patterns of usage may impart colloquial terminology to oil & gas properties, the foundation of petroleum property ownership is derived from real estate appraisal, which in turn is based in centuries of common law augmented by legal wrangling and subsequent definition.

## Property Ownership Interests – A Primer

A property that consists of both the surface real estate and the subsurface mineral rights is said to be owned in **fee simple**. A fee simple estate implies absolute ownership unencumbered by any other interest or estate. The fee simple owner has unrestricted access to and use of the surface land for farming, ranching, development, or any other legal use. The fee simple owner also has access to the subsurface to obtain water and exploit mineral deposits. In real estate practice this is often referred to as a "bundle of rights."[3]

Depending upon location and on the history of property rights in a region, surface real estate is commonly described as part of a property down to a certain depth (~100 ft.) from the surface. Mineral rights are usually defined as the real property below 100 ft. from surface. An important distinction is made between **real estate** and **real property**, where real estate is the physical land and appurtenances attached to the land (e.g., structures and facilities) whereas real property includes all interests, benefits and rights inherent in the ownership of physical real estate.

Mineral rights allow the owner of those rights to access, develop, and use any substance including water, oil & gas, hard minerals, sand and gravel, etc. that may be found. In most jurisdictions, and for general evaluation purposes, mineral rights are inclusive of oil & gas rights. There are exceptions, such as in Pennsylvania, where oil & gas rights are separate from other mineral rights. While not usually recognized as an evaluation issue, the United States is unique in that private persons can hold ownership of both the surface real estate AND the mineral rights underlying that surface property. In virtually all other countries, including Canada and Mexico, with some modest exceptions, the mineral rights are retained by the government. Private property ownership is a good thing but can be viewed as adding complexity to an evaluation of a property.

Mineral rights can be severed, in whole or in part, from the surface estate by sale or by assignment of those rights to another party, resulting in separate ownership of surface and subsurface rights. Subsurface rights that have been severed can be developed by the mineral rights owner either in agreement with the surface owner or through an easement.

Whether the oil & gas rights are owned in fee or are severed from the surface, these rights can be leased under a variety of terms to a third party for the purpose of exploitation. In the case of a lease, ownership of the mineral rights is retained by the Lessor; the Lessee obtains only a **leasehold right** to access, develop, and then to produce and sell oil, gas and/or other products from the property. Over many decades of experience, oil companies and others have experimented with the terminology of leasing mineral rights. This experience has led to what could be called a "**Standard Lease Form**." Not all leasing forms are identical; each prospective Lessee may have a form that differs in language or other respects from that of another company, or local law or custom may require variation, but the general context is the same. In addition, the experience of leasing companies has resulted in the lease language being very comprehensive and designed to cover all situations that have been previously encountered.

The common forms of an oil & gas lease include certain terms. The lease:

1. grants the Lessee access to the property with a degree of surface land use (easement) for roads and other facilities,
2. requires the Lessee to pay a royalty to the Lessor, usually in the form of a percentage of revenue from product sales,
3. is limited to a "primary term" of 5 to 10 years pending the start of production,
4. defines the terms of payment of any royalty or other payment to the Lessor,
5. allows the Lessee to assign and/or unitize all or part of the leased property,

6. specifies payment of "delay rentals" until production starts, and
7. provides that, once production begins, the Lessee may retain the lease for so long as production is obtained in **paying quantities**.

Paying quantities is generally defined to mean that the Lessee is producing the property at a profit. Additional terms may be imposed depending on the location of the property, the circumstances expected to occur during development and production, and the sophistication of the Lessor. These terms and others may have an influence on the valuation of the oil & gas rights, particularly as they affect the Lessor's interest and income. [see Valuation of Royalty Ownership Interests]

The Standard Form may be accepted by some Lessors but, in reality, it is only a starting point; particularly in competitive circumstances, all the terms are negotiable. The terms of a lease on one property may differ from a lease on another nearby property depending upon the experience and creativity of both the Lessor and Lessee. The development of natural gas in the Marcellus Shale in north-central Pennsylvania is a good example of the progression of leasing terms. When leasing companies began talking to land and mineral rights owners during the period from 2006 to 2008, the idea of oil & gas potential in the area was novel; many property owners accepted the Standard Form with few, if any, changes. The terms were a 5-year primary term, a basic royalty of 12.5% with no signing bonus, and generally unrestricted access to the surface for drilling, development, and for transportation of gas across the property or use of the property for gas storage.

As leasing activity increased, some landowners hired attorneys or other knowledgeable persons to help negotiate the terms of the lease. This led to increases in royalty percentages (up to 20% in a few cases), large signing bonuses (up to $5,000 per acre), restrictions on surface access, disallowance of any deductions from the royalty, no property usage for gas storage, and many other changes that, in some instances, made the addendum or amendments to the Standard Form lease more voluminous than the original document. From an evaluation standpoint, this progression of knowledge and ability to negotiate lease terms requires that an evaluator become knowledgeable of, and to carefully consider, the lease terms in any evaluation.

The conclusion of a mineral rights lease and the transfer of some, but not all mineral rights in a property creates a **leased fee estate** which is an ownership interest held by the Lessor with the right of use conveyed by lease to another party (the Lessee). The rights of the Lessor and Lessee are specified by the contract terms contained in the written lease agreement. The Lessee is said to hold a **leasehold estate** subject to the terms and ownership conditions of the lease.

The interests retained by Lessor and Lessee are referred to as "ownership." Mineral rights can be entirely owned (100%) by one party, or the property ownership can be fragmented among a number of owners. Ownership in a property can be broadly divided into two classes: Working interests and Royalty interests.

In industry terminology, the leaseholder or lessee is usually known as the **Working Interest Owner**. The owner of the working interest has the exclusive right to exploit the oil & gas on the property. The mineral rights owner, the Lessor, is generally known as the **Royalty Interest Owner** because he receives a payment or "royalty" from the Lessee.

The term **Working Interest ("WI")** derives from the fact that the Lessee does the development work, makes the investments, pays the operating costs, and arranges for the production and sale of the oil, gas, and any other products. In return, the WI owner receives the large majority of the revenue. There may be more than one WI owner if the Lessee has partners.

The **Royalty Interest ("RI")**, the interest retained by the Lessor, owner receives a fractional share or a percentage of the revenue but pays no costs and makes no investment except as may be stipulated by the leasing agreement. The royalty is usually paid directly out of sales, often by the oil and/or gas purchaser.

The share of revenue retained by the WI after satisfaction of all royalty and any other non-operating interests, is commonly referred to as the **net revenue interest (NRI)**.

As an example, assume there is a single WI owner, the WI is 100% or 8/8 and the royalty is 12.5% or one-eighth, then:
(NRI) = WI-RI
NRI = 100.0% - 12.5% = 87.5%

In any property or project there can be multiple WI owners and RI owners with varying shares of the WI and RI. If there is more than one WI owner, the WI and NRI will generally be divided proportionately. Where Lessee (A) executes a lease from the Lessor bearing a royalty of 12.5% and (A) thereafter assigns one-fourth (25%) of the WI to (B), the relative ownership interests are:

NRI (A) = (100 - 12.5%) * 0.75 = 65.625%
NRI (B) = (100 - 12.5%) * 0.25 = 21.875%

There is no change to the royalty interest.

In some circumstances, a non-operating WI owner (or in rare cases, a royalty owner) may elect to take all or a part of their share as an actual volume of oil or gas rather than receive a payment from the sale of their share. This is known as taking **in-kind**. Taking in-kind is not common but could occur when there are multiple WI owners and one owner has a more advantageous product sales option. In the evaluation of a property where one or more interest owners take their shares in-kind, the evaluation may have to account for and combine two or more income projections: one for the in-kind production and one for the balance of the property production. The royalty share remains the same whether the oil is taken in-kind or not.

The working ownership interests may change during the expected life of production. This situation often occurs in new drilling projects when a WI owner, typically a third-party investor, receives a certain WI share until the investment is recovered or "paid out," at which time the investor's WI share reverts to a reduced percentage and all other WI owners receive increased interests. A reversionary interest may occur when a certain cumulative volume of production is obtained, but more commonly occurs when a predetermined dollar amount of net income stream has been obtained by the investor. Such arrangements can take many forms. It is not uncommon for reversion to be based on a multiple of the investment or a specified volume of production, or, less commonly, a period of time.

As an example, assume that the operator (A) of a property wants to drill a new well and an investor (B) agrees to provide the $1,000,000 needed to drill the well. In return for the investment, (B) receives a 75% WI until the $1,000,000 (or a multiple thereof) is recovered from net revenue after all operating costs, royalties, and other obligations are met. At that point (payout of investment), the investor's interest changes or "reverts" to a 25% WI. If the property is subject to a 12.5% Lessor royalty interest, the respective Net Revenue Interest (NRI) after reversion is as follows:

Original Investor NRI = 75.0% * 87.5% = 65.625

Reversion Investor NRI = 25.0% * 87.5% = 21.875

# Pooling and Unitization

A special case of hydrocarbon property ownership involves the **unitization** or **pooling** of multiple properties and owners, both WI and RI, into a single production entity, or unit. Unitization is important where there is separate ownership of portions of the rights in a common field or reservoir. Unitization can make it economically feasible to conduct pressure maintenance, enhanced recovery operations, or conduct exploration activities.[4] Units can be created as development, production, and operational entities which are codified by legal agreements among multiple operators and mineral rights owners. For discussion, development, production, and operational units are collectively referred to as production units. Another form of unit is the less formal pooling unit which commonly refers to *"the bringing together of small tracts sufficient for the granting of a well permit under applicable spacing rules. Pooling is important in the prevention of drilling of unnecessary and uneconomic wells, which could result in physical and economic waste."*[5,6]

## Production Units

Combine properties primarily for the purpose of water-flooding, steam injection, or other enhanced oil recovery (EOR) operations, where production and injection can be expected to cross property lines, or where geology, geography, or operational circumstances suggest that a combination of properties would improve production and operating efficiency. Historically, production units have been some of the largest producing entities sometimes encompassing entire fields.

The formation of a production unit is a contractual arrangement that may include an Operating Agreement and a Unit Agreement. The **Operating Agreement** is a contract designating one or more companies as the Unit Operator and specifying the obligations of the Unit Operator and the participating parties. The **Unit Agreement** includes the description and ownership of the properties being assembled into the unit and the allocation of shares of the unit to each working and/or royalty interest owner.

In the formation of a typical EOR unit, each participating property is assigned a proportional share of the unit production, operating costs, investment, and income based on agreed-upon parameters, such as estimated remaining oil-in-place or remaining potential recovery on each property as a percentage of the total anticipated Unit oil-in-place or estimated recovery.

As an example, assume that fifteen companies own WI in ten properties which are being formed into a production unit. A geologic/reservoir/production engineering study determines that the ten properties have differing volumes of estimated remaining recoverable oil. The Unit Agreement may allocate shares in the Unit to each company based on the ratio of the recoverable oil on the company's property to the agreed upon recoverable oil for the entire Unit. Assume that the study determines that Property A has remaining recoverable oil of 5 million barrels and the total for the whole unit is 60 million barrels; Property A then has a one-twelfth or 8.3333% share of the total unit remaining recoverable oil.

The share assigned to Property A, often termed a **Tract Factor**, becomes the ownership basis for the value of the properties which form the Unit and for the owner of an interest in that property. Further, assume that a value is required for the XYZ Oil Company which is the owner of a 75.0% WI share of Tract A of the Unit. The value of the XYZ Company interest is derived from 75.0% of the Property A Tract Factor (8.3333%) of the Unit or 6.25% of total Unit production and net income.

Unitization of the ten properties does not change the royalty interest in the individual properties. The Royalty Owners in each property are incorporated into the unit through the unit agreement based on the terms of the respective lease agreements.

As an example, assume that a Lessor has granted a lease on Property A with XYZ Oil Company bearing a 12.5% royalty. The Lessor share of total unit production and sales would be:

RI (12.5%) *XYZ WI (75.0%) * Tract Factor (8.3333%) = 0.78125%

## Pooling Units

Less complex agreements usually involving a single WI operator and several oil & gas rights owners. A pooling unit is formed by the producing company to consolidate parcels of oil & gas rights on which the company holds leases into a single unit that can then be developed by drilling one well rather than a well on each property. Pooling units are most common in rural areas where small farms are found but can also occur in urban areas to combine residential and commercial parcels.

> As an example, leasing of oil & gas rights by companies in the Marcellus Shale in Pennsylvania often include many farm and small residential properties ranging from a few acres to several hundred acres, along with some large (several thousand-acre) hunting and recreational properties.

Drilling and fracturing of horizontal wells commonly require a minimum of 640 acres to accommodate a drill path along with a considerable amount of acreage for a multi-well drill pad. In order to meet that requirement and to avoid crossing any more lease lines than necessary, operators form pooling units of as many contiguous oil & gas rights properties as is practical to accommodate future drilling. Pooling units require approval of all affected oil & gas rights owners (WI and RI). The pooling "declarations," are filed as legal documents in the county where the unit is located. The declarations define the total mineral rights acreage ownership in the pool and allocate shares in the pool based on each interest owner's percentage of the total pooled acreage. Pooling Units do not require specific Unit and/or Operating Agreements; the terms of the pooling are usually defined in the declaration.

As with production units, interest owners in pooling Units are allocated a share of the unit usually based on mineral rights acreage. The *Black Unit* established by *Chief Oil and Gas* by the Declaration of Pooling and Unitization Agreement (Exhibit 1) effective May 20, 2009 included 715.4396 acres in Bradford County, PA. The pooling unit comprised 24 separate properties which were designated as "tracts." Tract 2 was the largest tract with 102.7730 acres in the unit and was allocated 14.365% of future Black Unit production and net revenue. The pooling unit combines the various oil & gas rights for development but does not alter or supersede those rights. For example, the royalty on Tract 2 may be 18% while the royalty on Tract 18 might be 12.5%; the rates do not change but are applied to the owner's share of the pooled unit.

ATTACHED TO AND MADE A PART OF THAT CERTATIN DECLARATION OF POOLING AND UNITIZATION AGREEMENT EFFECTIVE MAY 20, 2009, COVERING THE 715.4396 ACRE BLACK UNIT LOCATED IN BRADFORD COUNTY, PENNSYLVANIA.

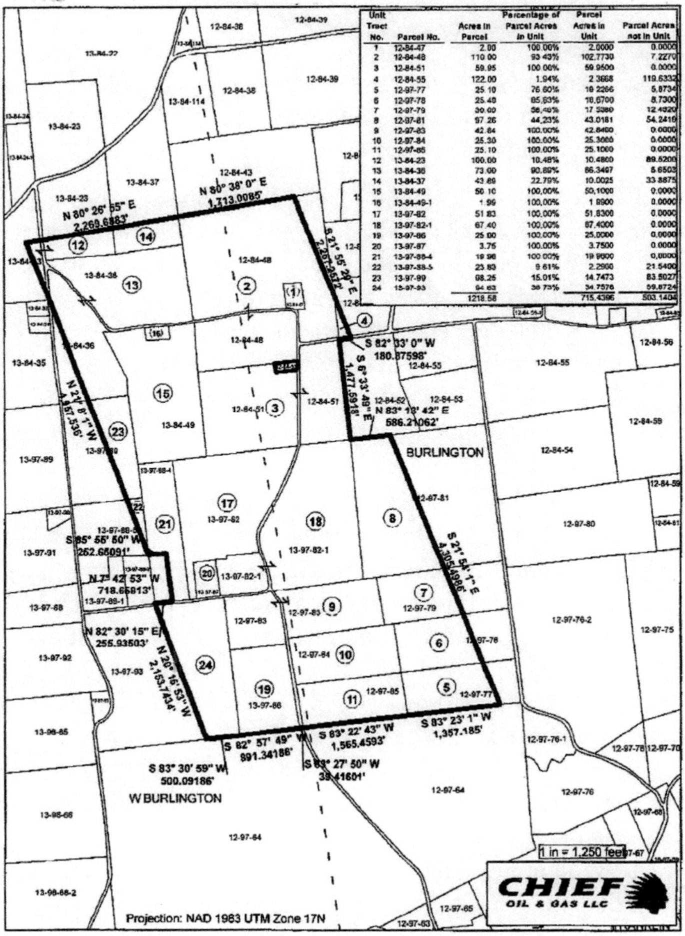


| Unit Tract No. | Parcel No. | Acres in Parcel | Percentage of Parcel Acres in Unit | Parcel Acres in Unit | Parcel Acres not in Unit |
|---|---|---|---|---|---|
| 1 | 12-84-47 | 2.00 | 100.00% | 2.0000 | 0.0000 |
| 2 | 12-84-48 | 110.00 | 93.43% | 102.7730 | 7.2270 |
| 3 | 12-84-51 | 59.95 | 100.00% | 59.9500 | 0.0000 |
| 4 | 12-84-55 | 122.00 | 1.94% | 2.3668 | 119.6332 |
| 5 | 12-97-77 | 25.10 | 76.60% | 19.2266 | 5.8734 |
| 6 | 12-97-78 | 25.40 | 65.63% | 16.6700 | 8.7300 |
| 7 | 12-97-79 | 30.00 | 58.46% | 17.5380 | 12.4920 |
| 8 | 12-97-81 | 97.26 | 44.23% | 43.0181 | 54.2419 |
| 9 | 12-97-83 | 42.84 | 100.00% | 42.8400 | 0.0000 |
| 10 | 12-97-84 | 25.30 | 100.00% | 25.3000 | 0.0000 |
| 11 | 12-97-85 | 25.10 | 100.00% | 25.1000 | 0.0000 |
| 12 | 13-84-23 | 100.00 | 10.48% | 10.4800 | 89.5200 |
| 13 | 13-84-36 | 73.00 | 90.89% | 66.3497 | 6.6503 |
| 14 | 13-84-37 | 43.89 | 22.79% | 10.0025 | 33.8875 |
| 15 | 13-84-49 | 50.10 | 100.00% | 50.1000 | 0.0000 |
| 16 | 13-84-49-1 | 1.99 | 100.00% | 1.9900 | 0.0000 |
| 17 | 13-97-82 | 51.83 | 100.00% | 51.8300 | 0.0000 |
| 18 | 13-97-82-1 | 67.40 | 100.00% | 67.4000 | 0.0000 |
| 19 | 13-97-86 | 25.00 | 100.00% | 25.0000 | 0.0000 |
| 20 | 13-97-87 | 3.75 | 100.00% | 3.7500 | 0.0000 |
| 21 | 13-97-88-4 | 19.96 | 100.00% | 19.9600 | 0.0000 |
| 22 | 13-97-88-5 | 23.83 | 9.61% | 2.2900 | 21.5400 |
| 23 | 13-97-99 | 98.25 | 15.01% | 14.7473 | 83.5027 |
| 24 | 13-97-93 | 94.63 | 36.73% | 34.7576 | 59.8724 |
| | | 1218.58 | | 715.4396 | 503.1404 |

**Exhibit 1.** *Chief Oil and Gas* by Declaration of Pooling and Unitization Agreement

# 4

# GENERALLY ACCEPTED VALUATION METHODS

An evaluation is done to estimate the economic benefit of a property as of a certain date. All valid and acceptable evaluations occur in the context of basic real estate valuation methods. As noted above, oil & gas properties are considered real estate in virtually all jurisdictions. The real estate aspect of oil & gas properties may have little obvious relation to the valuation of a property; however, there are at least two reasons to be aware of the influence of the real estate industry on valuation practice. Oil & gas evaluation borrows many concepts and procedures from real estate, which is where most appraisal techniques were developed.

1. First, the purpose of the evaluation and the jurisdiction in which the property occurs can influence the evaluation method that is applied. The term **jurisdiction** has been used because when an evaluation is being done for internal purposes, the internal control sets the rules of evaluation. However, if the evaluation is done for any regulatory purpose, such as ad valorem tax or estate tax, the rules are set by regulatory agencies and, possibly, by courts of law. These jurisdictions often require that evaluations follow, in form if not in function, real estate evaluation practices.
2. Second, it is not uncommon in petroleum literature, some texts, and in public and casual discussion to reference and commingle oil & gas evaluation terms and methods with real estate terms and methods, (usually unknowingly) resulting in unnecessary confusion.

Real estate appraisals are expected to adhere to specific guidelines and are often part of required evaluations for estate tax and other purposes; the reference texts and journals are about real estate – land, homes, buildings. Oil & gas evaluation developed separately – from the ground up, so to speak – and can stand alone without reference to real estate practice. For practitioners in oil property evaluation, the concepts common to real estate may seem an exotic terrane stuck on the continental plate of oil & gas evaluation. However, it may be useful to refer to real estate methods or at least understand how they might relate to oil & gas.

Generally accepted appraisal practice recognizes three approaches to estimate the value of real property. According to The Appraisal of Real Estate (ARE): *"Appraisers estimate property value with specific appraisal procedures which reflect three distinct methods of data analysis – cost, sales comparison, and income capitalization. One or more of these approaches are used in all estimations of value; the approaches employed depend on the type of property, the use of the appraisal, and the quality and quantity of the data available for analysis.*"[7] The potential application of these approaches to petroleum property evaluation requires discussion of the methods.

## Cost Approach

The **Cost Approach** estimates value based on the cost to replace or reproduce a structure or facility such as an office building, factory, HVAC unit, or other equipment. Essentially, this method estimates the value of a 20-year-old factory building by determining the cost to reproduce the building, as of the evaluation date, using current construction expenses, building codes, and other requirements. Similarly, the approach could value a stand-alone facility, such as a power generation plant, by estimating the cost to replace it with current equipment and construction requirements.

## Comparable Sales Approach

The **Comparable Sales Approach** relies upon the process in which a value estimate is derived by analyzing the market for similar properties that have been recently sold (the comparables) and comparing these properties to the property being valued (the subject). This approach is commonly used in real estate appraisal for residences, land, farms, and commercial/industrial buildings. It is simple in concept and is usually thought to give a good estimate of value for the appropriate types of property. The process has several steps. The appraiser researches the market for sales of properties with characteristics similar to the subject property. After verifying the data regarding both the property and the sale, the appraiser defines the similarities and differences between the comparable and subject property and then makes adjustments to the value of the comparables to match the attributes of the subject property.

This method works well for the types of properties to which it is normally applied. In residential real estate, all the necessary data is relatively easy to obtain from public records and from realtors' listings and databases. Banks and other sources provide financing and market information. If necessary, the appraiser can go to each property and accumulate data directly – measure the house and the rooms, check the heating, the roof, the plumbing, etc. He can talk to the buyer and the seller, real estate agents, finance company, and escrow agents. For further discussion, refer to Appendix A, Comparable Sales Method of Appraisal and Petroleum Properties.

# Income Approach

In real estate, the **Income Approach** is based on estimating the value of a property by capitalizing the revenue expected from the property. Real estate appraisal uses a method call **Direct Capitalization** which assumes a fixed level of income from the property and uses a capitalization rate to convert the income to a value. This method is usually applied to properties which produce rents to the owner such as apartment buildings, office buildings, leased factories and facilities, and, occasionally, large farms and livestock-producing facilities.

Real estate appraisers are expected to consider and apply all three methods in deriving a property value, for two reasons.

1. First, the type of property and data available may not be amenable to evaluation using just one method.
2. Second, using more than one method may account for characteristics and value considerations which might be missed using only one method and/or balance value considerations that may be under- or over-appreciated by one of the other methods.

As an example, consider a 10-year-old apartment building:

- An estimate of value could be based on the cost of reproducing or replacing the structure.
- An estimate could be based on obtaining data from recent sales of similar apartment buildings, making appropriate adjustments, and applying that value to the subject property.
- An estimate of value could be based on the expected income to be earned from rents assuming continued operation of the property.

# Application to Oil & Gas Property Evaluation

The **Cost Approach** has no application to oil & gas property evaluation because the property cannot be reproduced or replaced by new construction regardless of investment. Each field and property is unique both in time and in terms of the development and operation of the property by the owner. A property that was developed in the 1950s or even the 1980s would likely not be developed in the same manner as it would today. However, the value of the property as of January 1, 2023, is based on current conditions, not on a possible alternate drilling and production plan. Further, the method of operating Property A might be different from Property B but the value of Property A is based on the operating methods in place and not the methods on Property B. Going even further, properties in the same field or reservoir can vary in the characteristics that influence value.

Important caveat: This is not to say that the Cost Approach could not be used to obtain values for certain components of a property, such as a stand-alone gas-treating system or a co-generation plant, the value of which could then be added to the producing property value.

The Comparable Sales Approach has very little use in the evaluation of oil & gas properties, for several reasons. The overriding difficulty is access to appropriate data. Unlike residential and commercial real estate, there is no common searchable database of transactions from which to draw information. As noted above, even for residential real estate, an appraiser must assemble and analyze an extensive amount of information in order to make proper comparisons and adjustments to the comparables before estimating a value.

For oil & gas properties, the only information that may be available through publications or hearsay is that a certain property was sold and (sometimes) the estimated purchase price. An evaluator may be able to obtain some information about production, number of wells, oil gravity, and other parameters from public sources but, unlike a residential property, where the appraiser can obtain information from a multiple-listing service, there is no such source for oil & gas properties.

Oil & gas properties have most commonly been valued using the Income Approach which estimates the present value of a future stream of income. This choice results from the recognition that the source of value in a petroleum property is the income stream to be received from the production and sale of oil & gas and the knowledge that it is the method most commonly employed by investors in and appraisers of oil- and gas-producing properties.

## Application of the Income Approach for Property Evaluation

The Income Approach as applied to the evaluation of oil & gas properties commonly takes one or both of two forms: the income multiple and/or the discounted cash flow. The income capitalization model from real estate is not used. The **income multiple** is a simple method that is particularly useful for non-operated interests with relatively low income (i.e., less than $5,000 per year). The method uses an average of the monthly income over a three-to-five-year period and multiplies that average by an appropriate number of months to estimate a value. [See Valuation of Royalty Interests]

The **discounted cash flow** method of evaluation is an extrapolation of the form of Income Approach used in real estate appraisal. It is more complex than the real estate application and requires a wider array of information. Ordinarily, a property that has a reliable record of oil and/or natural gas production and sales should not be difficult to value by the Income Approach.

The utility of the discounted cash flow approach is that it expressly recognizes the income-producing nature of a petroleum property and provides a means for evaluation of the property in as much detail as desired, subject to data availability, taking into account the characteristics of the property and the evaluator's concept of future performance and economics. Every discernible physical or economic factor is related. A change in the input data or in an assumption used in one part of the income stream may affect a related component and require further change at another point in the model. The proliferation of computer-based economic valuation models permits an evaluator to analyze the subject property under a variety of current and future production/economic conditions. The end result is an estimate of a future stream of income that is unique to the property and specific to the date of evaluation.

## The Characteristics of the Property

A properly constructed production and income projection, whether of one well, a multi-well lease, or an entire field, contains all the definable and quantifiable characteristics of that property. The property characteristics and resulting projections may not be unique, but the property can be fully represented, nonetheless.

Assume, for example, a multi-well lease with sufficient production and operating history to allow the property to be analyzed without having to rely on data from other properties. The production performance of the property is a direct function of the physical nature of the reservoir rocks, the composition of the reservoir fluids, and the method of operation of the property. An evaluator must try to determine how the geologic characteristics of the property may affect both the recoverable volume of oil in the reservoir and how the characteristics of the reservoir may affect the ability of the property to produce oil and/or gas and the rate(s) of production. The wells will produce only what the rock permeability, porosity, etc., will allow; reservoirs with high water or gas saturations may produce more water or gas than reservoirs that are not highly saturated. Wells that are fracked, pumped, steamed, or produced intermittently will have those conditions reflected in the production history, operating costs, and other data. Therefore, if the projection of future production is based on past performance, the projection necessarily includes those same characteristics. Likewise, a proposed or anticipated change in future operations, presumably based on an analysis of the effects of such a change, would also reflect the character of the property.

Additional elements of the projection incorporate other characteristics. The **oil price** directly reflects the oil gravity and hydrocarbon composition of the crude. The **natural gas price** reflects heating value, transportation costs, etc. Ownership interests and royalties are property-specific. **Production taxes** are a direct function of location. **Operating costs** are a function of depth of production, number of wells, production methods, gross fluid volume, oil composition and gravity, location, and virtually every other characteristic of the property. The contribution of each characteristic may not be definable or quantifiable, and the influence of one may be partially offset by another, but there can be no argument that operating costs reflect the character of the property – and that they are not readily transferable except in broad terms.

An evaluator must also define the **operating system** that is necessary to obtain the production. The operating system will be determined, in part, by the physical characteristics of the hydrocarbon source, but will also be related to the business plan for the property as implemented by an owner/operator. The choice of operating system will influence the costs of production, such as the costs of lifting, gathering, and processing produced fluids. Such costs may include capital investments for wells and surface equipment plus other expenditures to meet regulatory requirements.

## Incorporation of Current and Future Economics

Construction of an income stream for a property allows the evaluator to incorporate not only existing economic conditions of the property but also to apply expectations for oil prices, gas prices, operating costs, taxes, and other factors. While the discussion of economics usually centers on future product price and operating costs, other areas such as the need for further investment, production or ad valorem tax increases, changes in royalty or other ownership, and federal and state regulation and income taxes, where applicable, must be considered. These expectations for the future are not trivial matters. Expectations are what drives investment decisions. Expectations also have influence on the remaining life and future production of a property by defining the economic limit of production and by invoking other economic conditions such as the cost of abandonment and clean-up.

## Reduction of Differences Among Properties

If constructed so that it reflects the characteristics of the property and includes reasonable economic expectations, the income stream can then render its greatest service to investors, owners, and evaluators by providing a means of directly comparing one property or investment to another on a quantifiable basis.

Consider, for example, two properties: Property A is a multi-well lease in Kern County producing steamflood oil from the South Belridge-Tulare Formation. Property B is a multi-well lease producing from the Seminole-San Andreas reservoir in western Texas. An evaluator knowledgeable of both properties could doubtless list a large number of characteristics that differ between the properties. However, at the same time, he or she would be hard pressed to find a characteristic that is not reflected by some part of the income stream.

The first thing someone might point out is the difference in oil gravity. The oil gravity is different, but the difference is reflected in the price paid for the oil along with the costs of production. The same thing is true for the production rates, decline rates, gas production and net gas sales, water production and disposal, production taxes, etc. Each item is included in the income stream based on how it relates to the specific property being valued – not to some other property. Even the pressures of environmental and other regulations are being factored into operating costs, as producers find ways to allocate these costs.

The result of constructing an income stream for Property A and for Property B are two projections of future income that reflect, explicitly and implicitly, all the salient characteristics of the properties and most of the more subtle ones as well. The income streams can be directly compared to each other. Intangible differences among properties may include restrictive environmental or regulatory constraints, particular benefits to be derived that are not entirely quantifiable (location of property, "strategic fit" with existing operations), or "management" decisions.

# The Discounted Cash Flow Model

The Discounted Cash Flow Model consists of six broad elements: the production schedule, product prices, ownership interests, costs of production, capital expenditures, and regulatory compliance. Any of these elements may be expanded and can be made as simple or as complex as necessary.

## Production Schedule

The starting point for any evaluation is an estimate of the future production of crude oil, natural gas, and water, plus any associated production such as natural gas liquids (NGL) or condensate. Reservoir and production engineering techniques allow evaluators to estimate potential recovery and to schedule future production of oil, gas, natural gas liquids, and/or other products to meet future operations, whether it be continued decline, new development, EOR, or a combination of production plans. The production schedule, if derived from an appropriate and reliable study and subjected to careful consideration of alternatives, can incorporate all of the reservoir, geologic, and operational characteristics of the property and the known or perceived risks associated with those characteristics. Extrapolation of historical production performance from the property (or a similar property) is the most direct method of estimating future production when either oil or non-associated gas is the primary production stream. Associated gas may be projected as a function of oil production using a fixed or variable Gas-Oil Ratio (GOR). Liquids extracted from produced gas may can be projected as a fixed or variable yield per volume of produced gas.

## Product Prices

The Product Prices of oil, gas, and NGL used in an economic evaluation may depend upon the purpose for the evaluation. Product prices vary depending on oil gravity and are subject to deductions for treating, transportation, diluent or other factors. Unless a particular price forecast is imposed by company guidelines, regulation, or third-party source, the estimate of future product price should be a function of experience and foresight.

## Ownership Interests

Ownership Interests specific to the property must be applied and, further, can be varied over the life of the property to incorporate any changes which may be expected to occur in the legal ownership of production and revenues.

## Operating Costs

Operating Costs used in valuing a property should be as detailed as the available data will allow. To the extent that the evaluator is able to identify potential cost areas of the property, the income approach allows all relevant costs to be calculated and considered in evaluating a property. The identification and quantification of operating costs can incorporate many of the conditions that make properties unique. Heavy oil properties requiring diluent, steam injection, or special chemical treating can have these costs included in the income stream. Properties that are high (or low) water producers may have that circumstance reflected in the lifting, treating, and disposal costs. The depth of production, requirements for maintenance or remedial work, and any other cost that can be identified and evaluated, can be made a part of the income stream calculation on the basis appropriate to the cost and/or property. Periodic cost items can be included on a time schedule. Severance- and production-related taxes as well as ad valorem taxes can be included as appropriate.

## Capital Investment

The development and operation of a petroleum property invariably requires some capital investment at the outset, whether for drilling wells, installing equipment, obtaining permits and approvals, or initiating production. Further capital investment during the life of the property could occur for expansion with new wells, replacement of aging equipment, or startup of an EOR project. There will also be required expenditures at the end of production for abandonment and remediation activities.

## Regulatory Compliance

Regardless of location and type of operation, oil- and gas-producing properties are subject to many differing forms of government regulation. Compliance with these regulations is an increasing and inescapable part of property evaluation. To the extent possible, the costs and operational impacts of reasonably anticipated regulatory compliance including abandonment, restoration, and remediation, should be included in an evaluation.

# 5

# PRODUCTION ESTIMATION AND SCHEDULING

**Economic value** is derived from the production and sale of oil, gas, and other hydrocarbon products. The estimation of the volume of producible hydrocarbons and the projection of a reasonable schedule for recovery of that volume is the most work-intensive part of any oil property evaluation. Reservoir analysis and production scheduling are a large part of the applied science of petroleum engineering developed over decades of field experience and research work. The procedures and techniques for estimating future production and potential recovery are the subject of numerous publications which document petroleum engineering methods in considerable detail. It serves no purpose to reproduce the content of those works in this study.

However, it is useful to review some of the methods of constructing projections of future production and the related issue of potential (ultimate) recovery from the property. For the purpose of most property evaluations, the primary objective is a rational projection of future production that can be translated into an estimate of future income and economic value: the Jim Smith Rule. The true ultimate recovery from a property is known only after production has ceased and the property has been abandoned. Prior to the conclusion of production, ultimate recovery is never more than an estimate.

## Actual Performance and Similar Properties

The most common type of property to be evaluated for either investment analysis or required valuation is one that is already in production and has an operational performance history that can accessed and interpreted. It is not unusual, in practice, to conduct evaluations of properties that have been producing for 10, 20, 50, or more years. Where sufficient documented history is available, the actual performance of the property should probably be considered the most reliable indicator of the future performance, provided that conditions remain essentially the same. Note that the **"documented history"** is more than a chronology of monthly or annual oil, gas, and water production from one or more wells. It also includes, where possible, a history (in broad terms) of the operation of the property such as the development sequence (wells drilled and when), remedial work, changes of operator, regulatory environment, and any other events that have influenced the performance of the property. As will be discussed later, the production history can be analyzed using numerous criteria to provide a rational projection of future production.

In the common case, historical performance data is available from the operator or owner of the property, Formats can vary and data is not always consistent; however, in most cases the information is in a readily usable form. Most operators can provide electronic tabulations of production, etc., but written reports can also be used. In those states that require reporting of oil & gas production along with ancillary data and where that information is available from a state agency, those sources should be accessed either in place of or, preferably, in conjunction with the data from the operator. In many respects, the choice of data source is a function of the complexity of the property and of the evaluation. In the evaluation of a single property or group of properties with a modest number of wells, using on-line data from the state may be the quickest and most efficient approach. However, evaluation of a large integrated property may require consultation with the operator.

> Another caveat: not all state agency databases are created (or managed) equally; some are more user-friendly than others. Commercial sources can be useful when available; however, the companies that maintain these databases usually obtain their information from either the state or, rarely, from the operators of the properties. Further, the data may be in a truncated format. The cost of obtaining data from a commercial source has to be balanced against the efficiency of access and quality of data.

Regardless of the source of data, consideration must be given to the extent of data that will be necessary. This means: How many years of prior history is sufficient to cover all the variables in the property performance and allow the development of a reasonable basis for projection of future production?

Another useful source of data is the historical performance of a **similar or analogous property** which can be shown to have the same general reservoir and operating conditions. Data from similar properties is particularly useful when the subject property has a short and/or unsettled production history. Assuming that data for a similar property can be accessed, it can be used to either simulate the subject property or supplement the information and analysis of the subject property.

> As an example, evaluations of new wells in the Marcellus Shale of north-central Pennsylvania can be done by integrating the performance of older wells into the analysis of the new wells to create production projections for the new wells.

> Continuing caveat: properties may be similar, but no two properties are exactly alike, care is necessary when using other properties as models. Common sources for similar property data are state agencies or commercial databases.

# Undeveloped, Underdeveloped, and/or Expanded Production Properties

When the property being evaluated has a documented performance history which can be analyzed, that actual performance approach should be used. Where there is little or no production history for the property and questionable analogy data, it may be necessary to estimate potential production using some of the concepts of reservoir engineering and production analysis. This essentially involves the estimation of potential recovery and working backwards. The property may have been subject to little or no development drilling and/or testing. Either circumstance may require sophisticated geologic analysis and reservoir engineering studies combined with computer modeling, and/or other approaches that provide information for the evaluation. However, it may also be necessary to anticipate that expanded production may be obtained through further drilling, changes in the operation of the wells, introduction of enhanced recovery, or other potential. The estimation of potential recovery could be broken down into four steps, each of which would narrow the final estimate.

Step One: Define the Reservoir Recovery Mechanism

Step Two: Estimate the Producible Volume of Oil-in-Place

Step Three: Determine the most efficient/effective Methods of Production

Step Four: Project the Form and Rate of Production

Steps One and Two can be accomplished separately but, as a practical matter, the producible volume and the reservoir production mechanism are inseparable. Solution gas drive results in a different production performance and ultimate recovery than does water drive. Gravity drainage is different from solution gas drive, and the production mechanism requires a different model. However, it is not always possible to accurately define the reservoir recovery mechanism, so estimation of producible volumes may require more than one model for estimating recovery. As a practical matter, the reservoir mechanism may usually be ascertained by reference to other fields in the region, a rational analysis of the available geologic/geophysical context, and the likely fluid content of the reservoir.

Step Three is interactive with Steps One and Two in the sense that the information gained from the analysis in those steps can lead to a decision about:

(a) whether to produce by flowing or artificial lift production,

(b) the development well density,

(c) whether to drill horizontal or straight-hole wells,

(d) when to introduce a method of production stimulation, or a myriad of other operating conditions.

This is an iterative process that will be repeated before a final plan is developed. The potential volume and production rates, derived as part of Step Four, should be the volume of potential recovery for most evaluation applications.

The necessary analysis may be accomplished through one or more of several sources:

- Investigation of the performance of analogous properties.
- Reservoir and production studies performed by the operator and/or by third parties.
- Original reservoir and production studies done by the evaluator.
- All of the above.

The use and analysis of data from similar properties was discussed above but, in this circumstance, may involve more than accessing and using production performance. Depending on the geologic, operating, and other data available, it may be possible to construct a similar property and extract reservoir information that could be applied to a model of the subject property. This approach can be especially applicable when the operator of the similar property is the same as the operator of the subject property. Analogous property performance can be determined by the evaluator with access to sufficient data and with enough experience to be able to evaluate the differences in performance among the properties relative to the subject property. Reservoir studies done by or for the operator almost always include some incorporation of data from analogous sources and can be helpful to an evaluator who has sufficient experience to assess the quality and utility of the work done and is able to integrate the methods and results into the valuation. Original reservoir studies done by the evaluator for the operator are not unusual for internal valuation or for some required valuation purposes, such as for ad valorem tax or eminent domain; however, for most other purposes, the cost and time required may be prohibitive.

The *All of the above* source is rare and potentially expensive but has the advantage of combining information from multiple sources, which may provide a wider range of understanding than a single source.

# Primary Production in Petroleum Reservoirs

Right about here, in most evaluation texts, the narrative transitions to a lengthy discussion of reservoir engineering and the various methods of estimating future recovery based on the geology of the reservoir, pressure and fluid flow characteristics, well design, and production development. As noted earlier, there are numerous texts that cover those subjects in great detail. In addition to petroleum engineering and geology books and papers, many companies have developed computer-based systems to evaluate oil & gas reservoirs including the estimation of future production; some of these are commercially available to evaluators. Given that those sources exist and can be readily accessed, this discussion will not repeat the information in any depth. Rather, in order to preserve the continuity of reservoir engineering to economic evaluation, a synopsis of basic methodologies, including the Volumetric and Material Balance methods of estimating recoverable oil & gas, is presented in Appendix B – Primary Production in Oil Reservoirs. The following discussion makes reference to and borrows from certain texts.

- *Petroleum Production Handbook*, Vol. II, Reservoir Engineering, edited by Thomas C. Frick, McGraw-Hill, 1962; hereinafter referenced as **PPH** with chapter and page noted as 37-10, respectively.
- *Petroleum Engineering Handbook*, edited by Howard B. Bradley, Society of Petroleum Engineers, Third Printing, 1987; hereinafter referenced as **PEH** with chapter and page noted as 37-10, respectively. PEH is largely a reprint of PPH with some new material.
- *Applied Petroleum Reservoir Engineering*, Craft, B.C. and Hawkins, M.F., 1959 (there are later editions), Prentice-Hall; hereinafter referenced as **C&H** with standard page numbering.
- *Petroleum Reservoir Engineering*, Amyx, James W., Bass, Daniel M., and Whiting, Robert L., 1960, McGraw-Hill; hereinafter referenced as **AB&W** with standard page numbering.

Each of the above sources exist in later editions where, for the most part, the only changes are the addition of color and the price of the book.

## Primary Production Performance – Oil Reservoirs

The Volumetric, Material Balance, and other methods of estimating potential production are (Appendix B) valuable tools but are often limited by a paucity of data; mathematical calculations that, in being simplified, leave out or assume certain reservoir conditions to be true; and the assumption of uniformity of conditions throughout the reservoir. In contrast, the **Production Performance** approach implicitly includes all reservoir and production operating conditions that would affect performance. When production is not curtailed by regulatory or other artificial conditions, the volume of production from a well is a direct result of the interaction, however great or small or uniformly dispersed, of all reservoir rock and fluid properties with the existing wellbore and operating conditions. Since oil reservoirs are finite in volume, production over time causes a reduction in pressure, which causes a decline in the rate of production.

## Fluid Flow in Porous Media

The foundation of petroleum reservoir and production engineering is simple: mobile fluids (oil, gas, water) will flow through porous media (i.e., unconsolidated sandstone, cemented sandstone, limestone, fractured shale) along a path (linear, radial, spherical) from Point A to Point B if the pressure at

Point A is greater than the pressure at Point B. The pressure can be artificial, such as in laboratory tests, or natural, such as in a petroleum reservoir. The pressure can result from:

- Expanding solution gas,
- Expanding free gas cap,
- Water influx (natural or artificial),
- Gravity drainage, or
- All of the above.

Everything in petroleum property evaluation, down to and including the income stream that defines the value of a property, derives from that simple set of facts. Having said that, literally hundreds of text books, professional papers, and college lectures have expanded the application of that concept, derived from Darcy's Law, into a wide range of equations, some more complex than others, to describe fluid flow in porous media under every conceivable situation. All of those illustrations start from the same place.

[Author's Note: A friend of a friend in college (and after) used to begin every technical analysis with the basic equation, F = ma, and expand from there to his conclusion. It was entertaining but, after a while, most of us learned to skip ahead to page 10.]

## Application of the Radial Flow Equation

Rather than spend a lot of time and space developing the relevant reservoir equations for all the possible circumstances that might be encountered in evaluating an oil- and gas-producing property (and thereby duplicating all the sources available elsewhere), one equation will suffice as an example and a springboard into primary production performance analysis. In the hypothetical situation where fluid flow occurs in a uniform, horizontal, radial plane under steady-state conditions (no radical changes in external or internal physical properties), the flow of fluid – in this case, oil – into a wellbore can be described as follows:

$$q = \frac{2\pi kh\left(p_e - p_w\right)}{\pi \ln\left({}^{r_e}\!/_{r_w}\right)} \qquad \text{[PPH 23-14 (20)]}$$

This simple but rather elegant equation relates the flow rate ($q$) to the thickness ($h$) and permeability ($k$) of the strata, to the radius of drainage ($r_e$) and the wellbore ($r_w$), as a function of the difference in pressure at the drainage boundary ($p_e$) to the pressure in the wellbore ($p_w$). Assuming that the components of the equation are known or can be reasonably estimated, a flow rate can be calculated at initial conditions and/or at later times as conditions change. The equation can be converted to calculate the flow rate of oil at surface conditions by adding an oil formation volume factor ($B_o$) and viscosity ($\mu$):

*"It is convenient in many applications of Darcy's law to introduce commonly used oil-field units. The following is a summary of the more common equations with the conversion factors to convert to oil-field terminology."* Arps in PPH 23-14

$$q = \frac{7.082kh\left(p_e - p_w\right)}{\pi B_o \ln\left(r_e / r_w\right)}, \text{ Bbl/ Day} \qquad \text{[PPH 23-14 (26)]}$$

$$q_w = \frac{39.76\left(kh\left(p_e - p_w\right)\right)}{\mu \ln\left(r_e / r_w\right)}, \text{ SCF/Day} \qquad \text{[PPH 23-14 (29), C\&H p.284 (6.24)]}$$

This solution is not unique: Steady-state conditions in uniform horizontal plane reservoirs, radial or otherwise, do not exist in nature. As a result, there are numerous variations of the radial flow equation which attempt to account for a range of actual or perceived differences in the nature of the fluid flow and the properties of the reservoir within the drainage area.

# Production Decline

The radial flow equation, in all its many variants, demonstrates that primary production of oil & gas is a direct function of the difference in pressure between the reservoir and the wellbore. Since oil & gas reservoirs are finite in volume, whether that volume is the physical extent of the reservoir or the volume within the drainage area of the wellbore, production over time causes a reduction in pressure within the reservoir which, in turn, results in a decline in the rate of production over time. This change in production rate can be calculated using a variety of equations depending on the perception of reservoirs' conditions but can also be represented and analyzed in graphical form.

The combination of time, production rate, and cumulative production (to the date of the evaluation) can be used to estimate both remaining potential production and productive life. When there has been sufficient history to establish a production trend for a property, the three variables may be plotted as graphs, commonly known as **decline curves**, which can be extrapolated to illustrate future production and estimate ultimate recovery.

The production decline rate is normally expressed as an annual rate for convenience of production projection and for comparison among properties. The decline curve may demonstrate one of three forms or shapes: **exponential**, essentially a straight line of constant slope, **hyperbolic**, a continuously flattening curve which can be described mathematically, and **harmonic**, a special case of the hyperbolic decline. The differences between and among the forms can be subtle, particularly in the later life of production, but distinguishing between the forms can often be important in estimating future performance and recovery.

Irrespective of form, the production decline ($D$) can be described in two ways: The **nominal decline rate** is the negative slope of the curve of the natural log of the production ratio (q) at time (t), or

$$D = -\frac{d\ln q}{dt} = \frac{dq/dt}{q}$$ [PPH 37-43(51), PEH 40-28(49)]

The **effective decline rate** is more common in actual practice and has the form of a loss rate:

$$D_e = \frac{q_i - q_1}{q_i}$$ [PPH 37-43(52), PEH 40-28(51)]

# Hyperbolic Decline

While not always obvious, most primary production occurs as hyperbolic decline. Hyperbolic decline may only become evident after some production has occurred and may require careful analysis to be discerned. In the late life of production, hyperbolic decline approaches asymptotic conditions and can then be approximated by a constant-rate decline. The method can be used to evaluate virtually all production histories, but experience in evaluating – gas and oil production from fractured shale reservoirs, such as the Marcellus and Haynesville, and – cyclic steam performance has shown the hyperbolic model to be particularly useful in analyzing the performance of the early life of that production.

For the hyperbolic decline, the decline rate can be approximated by:

$$D = -\frac{\left(dq/dt\right)}{q} = bq^n$$ [PPH 37-44(58), PEH 40-28(54)]

where ($n$) is a fractional power of the production rate between 0.0 and 1.0; it is a function of the reservoir and fluid conditions as reflected in the production rate while ($b$) is a constant derived from initial conditions:

$$b = \frac{D_i}{\left(q_i\right)^n}$$ [PPH 37-44(59), PEH 40-28(55)]

The production rate (q) at a later time (t) is a function of the initial decline ($D_i$):

$$q = q_i\left(1 + D_i t\right)^{\frac{-1}{n}}$$ [PPH 37-44 (60), PEH40-28(56)]

The cumulative production ($N_p$) is calculated as:

$$N_p = \frac{(q_i)^n \left((q_i)^{1-n} - q^{1-n}\right)}{(1-n)D_i}$$ PPH 37-44 (61), PEH 40-28(57)]

Most rate-time decline curves are hyperbolic with values of (n) between 0.0 and 0.7, while the majority are between 0.0 and 0.4. Published tables relating time to loss rate and ($N_p$) for various decline rates may be useful. [Arps, PPH 37-46] However, development and production of tight gas and oil-shale reservoirs using the combination of horizontal drilling and hydraulic fracturing has provided information from many formations (i.e., Barnett, Marcellus, Eagle Ford) that production not only follows a hyperbolic form but often demonstrates values for (*n*) that approach and, not uncommonly, exceed 1.0.

## Exponential Decline

Exponential Decline is often thought of as constant-percentage decline where the nominal rate is:

$$D = -\frac{d_q / d_1}{q}$$ [PPH 37-43(53), PEH 40-28(49)]

after integration,

$$q = q_i e^{-Dt}$$ [PPH 37-43(54), PEH 40-28(50)]

where (t) is the time period over which (q) is to be calculated and to which (D) must conform. Further integration yields:

$$N_p = \frac{q_i - q}{D}$$ [PPH 37-43(55), PEH 40-28(51)]

where ($N_p$) is the cumulative production during the period between ($q_i$) and (q). If $q_a$ at abandonment is substituted for (q), then ($N_p$) is the ultimate recovery at decline rate, (D). In practice, unless there is a reason to be sure that production will continue to follow a hyperbolic form, production is commonly extrapolated as exponential; the use of a straight-line projection is convenient and is easily applied to graphical representation of production rates versus time (date). Production from older properties with decades of history are commonly estimated from exponential decline models.

## Harmonic Decline

Harmonic Decline is a special and rare case of hyperbolic decline where n is usually found to be 1.0 and where:

$$D = -\frac{dq/dt}{q} = bq$$ [PPH 37-44(66), PEH 40-29(62)]

in which,

$$b = \frac{D_i}{q_i}$$ [PPH 37-43(67), PEH 40-29(63)]

and, after integrating,

$$q = \frac{q_i}{1 + D_i t}$$ [PPH 37-43(68), PEH 40-29(64)]

and

$$N_p = \frac{q_i}{D_i} \ln \frac{q_i}{q} = \frac{q_i}{D_t} \ln r$$ [PPH 37-43(69), PEH 40-29(65)]

The Nominal decline form ($D_n$) can be related to the Effective decline form ($D_e$).

Exponential: $D_e = 1 - e^{-D}$ [PPH 37-45(73), PEH 40-29(68)]

$$D_n = -\ln\left(1 - D_e\right)$$ [PPH 37-45(72), PEH 40-29(69)]

Hyperbolic: $D_{ei} = 1 - \left(1 + nD_i\right)^{\frac{-1}{n}}$ [PPH 37-45(75), PEH 40-29(72)]

$$D_{ni} = \frac{1}{n}\left[\left(1 - d_{ei}\right)^{-n} - 1\right]$$ [PPH 37-45(74), PEH 40-29(71)]

Harmonic: $D_{ei} = \frac{D_i}{1 + D_i}$ [PPH 37-45(77), PEH 40-29(73)]

$$D_{ni} = \frac{D_{ei}}{1 - D_{ei}}$$ [PPH 37-45(76), PEH 40-29(72)]

When extrapolation of historical performance may not be appropriate, such as when projecting future production from a new field or reservoir or for an EOR project, production estimates can be obtained by converting the volume results of the material balance, frontal advance, or other method into annual or monthly production. This may require identifying a limiting condition, such as lifting capacity or injection rates and back-calculating a production rate. However it is done, an evaluator must attempt to compare the production schedule with other similar projects or fields known to exhibit good reservoir and operating practices. Also, the selection of a schedule may have economic impacts, such as the requirement for investment, increased operating costs, or in some circumstances, royalties and/or taxes which may cause an alteration of the selected production schedule. Finally, as noted earlier, the proper analysis of the source and form of the production data is necessary to a valid income stream evaluation.

# Production Performance Analysis

The progression of recovery estimation methods from early-stage volumetric or material balance calculations (Appendix B) to historical performance analysis provides increasing accuracy and reliability. In practice, the methods can be shown to be complementary over time so that successive use of the methods may be used as a check on previous efforts. For primary recovery estimation, examination of historical production performance is the most reliable method assuming that sufficiently reliable data is available and the data is properly analyzed. Production performance has two mutually supporting objectives: first, the derivation of a future production schedule for a property, and second, the estimation of the ultimate recovery from that property. Performance analysis can be used for a single well, a select group of wells, a property, or an entire field.

## Production Information

Use of the production performance methods of evaluation, either alone or in combination with other methods, requires the collection and analysis of historical and current data to define a trend that can be extrapolated as a production schedule. There are numerous sources of data including, but not limited to: operating company production records, oil & gas sales records, individual well tests, reports to and databases of regulatory agencies, and public or commercial data-gathering firms.

The operator of a property is generally the best source of production and operating information. Assuming a good rapport is established, the operator/owner can usually provide production data in multiple formats by well, as aggregations of data from multiple wells, and for various time periods. The operator can also combine production data with information:

- Individual well records,
- Property operating changes,
- Testing and reporting methods,
- The procedure for allocation of production among wells,
- Production treating and transportation,
- Sales options,
- Regulatory controls influences, and
- Potential changes in operations, among much other information.

That said, the degree of cooperation of the operator is often inversely related to the purpose of the evaluation and the interest of the operator in the results. An evaluation done for investment decision making may be fully supported, but, if the evaluation is for another interest owner, such as a royalty owner, the operator may only provide limited cooperation. Regardless of the degree of support, access to production data can be problematic. Operating companies often employ production data collection, recording, and reporting systems that serve their purposes but which do not translate into formats that can be used to construct performance histories for wells and properties. The data may have to be converted to a more useable format which could be time-consuming, and the depth of the history may be limited. Accessible records may only go back a few years or may not include all production (such as gas used on the property as fuel or water production). Even so, starting with the operator and employing friendly persuasion is recommended.

Access to the production and other information regarding individual wells, properties, and ongoing projects has expanded exponentially over the past 20-30 years, primarily as a benefit from internet access. All major petroleum-producing states require reporting of oil, gas, and (usually) water production and have used that data to compile extensive collections of production and other information in digital format. For the most part these data collections have been made available to the public in comparatively user-friendly website formats. Some of these systems are more sophisticated than others, but, in general, collected data is reported on a well-by-well basis. This approach can be difficult to manage, particularly if a property involves a large number of wells (think Midway-Sunset with hundreds of wells) that may span decades, and may include production from different reservoirs and employing changing recovery methods.

Accessing, collating, and manipulating data from these sources can be time-consuming and tedious. Some sites allow aggregation of data; others do not. The sites often also provide access to:

- location maps,
- general geology, and
- well histories, such as drilling and workover events, well testing, and regulatory actions.

The ability to obtain data in detail quickly, usually in a readily usable format (.cvs or. xls) and at little, if any, cost has made practical the efficient evaluation of properties that was not possible in the past.

Similar data bases are maintained by commercial sources which obtain data from the state agency collections and offer access to the online data for a price. Probably the most useful aspect of the commercial sources is one-stop shopping convenience; the data from a number of state agencies may be available on one website in a standard format which reduces the need to access individual state data sites with possibly differing formats. Some of these sources are more efficient and economical than others, but the ability to access and use this data for a price represents an advantage to evaluation practice.

When available, accessing more than one source is a good idea. A combination of data from the operator along with data from public or commercial sources has the advantage of:

- incorporating the operator's experience in managing a property,
- allowing the opportunity to question the operator about changes or anomalies in production data obtained from public/commercial sources, and
- providing an opportunity to gain a more detailed overview of the operating history of the property.

One issue that often needs to be addressed is the method of testing and recording production for internal purposes and reporting to local/state agencies. Unless individual well production is accurately metered at the wellhead and/or subjected to rigorous and regular testing, some form of **allocation** is commonly employed which may or may not reflect the actual well performance. Aggregation of production at the point of collection for the property is necessary.

The **reluctant operator conundrum** can often be at least partially resolved by asking for the American Petroleum Institute (API) numbers of the wells on the property being evaluated and then going to the government or commercial data bases and extracting the needed information there. This approach has proved particularly effective when working with smaller operators. Experience has shown that obtaining a list of API numbers is relatively simple. If followed by obtaining the production data from another source, the operator may then be willing to answer specific questions rather than have to supply volumes of data.

## Verification of Data

Finally, a reminder that all data, regardless of source, must be carefully reviewed for rationality and to define the reason for any variations in data from one source to another. An example is the need to equate production during short production periods (February) and downtime (during pump changes or other well work) to production during "normal" periods. Changes in production caused by mechanical alterations, such as opening or closing flow valves, pump changes, and/or pump speeds, must be normalized and may require additional graphical analysis. Other changes, such as restrictions of product sales to pipelines, reductions in operations due to price fluctuations, and other factors, must be considered and accounted for in any production estimates. The most common variations in performance are those caused by:

(a) short producing periods,
(b) mechanical changes,
(c) regulatory restrictions, and/or
(d) capacity restrictions. Review production data for:

- Duration of Producing Period
  - Short months (February)
  - Wells produced less than full month
  - Production wells on timers or other controllers which can imply constant (flat) rates
- Mechanical Changes
  - Requires working with field personnel
  - Production reductions due to well downtime for repairs, etc.
  - Increased or (rarely) diminished production due to change in flow valves, pump size, stroke length, or pump speed
  - Increased or (rarely) diminished production due to down-hole well workovers
  - Decline in production due to increasing fluid levels, etc.
- Regulatory Restrictions
  - Proration (regulatory or contractual) – limitations on production rate and/or duration of production within a certain time period, usually a month
  - Limits on non-marketed gas disposal – flaring and restrictions on flaring of gas may inhibit oil production, particularly in high gas-oil ratio (GOR) wells
  - Limits on wastewater disposal – controls on the volume of produced water that can be disposed of or used for enhanced recovery projects could inhibit oil production, particularly in high water-oil ratio (WOR) wells
  - Air emission requirements – restrictions by state or local agencies on undesirable air emissions from production facilities could result in wells and properties being shut-in for days or weeks during periods of air pollution
- Capacity Restrictions
  - Limited capacity of surface storage and/or shipping facilities and/or gathering system or regional pipeline access. Particularly common in new development areas
  - Contract Volumes – limits on the volumes that can be sold from a property due to transportation or other constraints.

## Graphical Methods – Decline Curves

The decline form equations can be used to calculate and project future production when reasonable estimates of the appropriate input data can be made. However, that approach requires a lot of arithmetic and can be tedious. The more common approach is to use available data to create various graphical representations of performance histories and then extrapolate actual production of oil, gas, and other useful information (water, production days and wells, and, where necessary, injection data), using comparatively simple graphical methods. The utility of graphical methods, when used appropriately, is that the need to involve complicated calculation methods is largely obviated; the graphical methods incorporate all or most of the conditions encountered in the many and various equations. Further, graphical analysis can be applied regardless of the shape of the decline. The following discussion of graphical methods is illustrated by examples of application to certain oil- and/or gas-producing properties in various locations.

The most common representation is the semi-log rate-time plot where the date of production is plotted as the independent variable (X) and production rate is plotted as the dependent variable (Y) where X is in months and Y is barrels produced (see below). This curve, with sufficient definition, can be extrapolated into the future to estimate future production and potential recovery. The only requirements for extrapolation are that the decline curve demonstrate uniformity of shape and that there be an endpoint. Leaving the endpoint consideration for now, the uniformity of shape is required to ensure that performance is the result of the interaction of reservoir and mechanical operating conditions and is not being altered or influenced by artificial conditions or restraints.

When there are sufficient data available and external effects are limited, this form of rate-time representation incorporates all the characteristics of the reservoir rock, the produced fluids, and the operating system in use. When the data extends over a long enough period of time to establish "stable" conditions, the rate-time performance will allow definition of the form of decline dictated by the production characteristics.

**Note:** There are numerous internal, proprietary, and commercial computer programs that can be used to create decline curves and analyze production performance. Some are more sophisticated than others but most can be effective in derivation of decline rate and other information. There is not space or desire to review those programs here; however, before incorporating any of these programs, it is important to determine how the programs reach their results. Many commonly used programs use a simple least-squares fit of data while others allow adjustment of data in accordance with the verification of data noted above. The following examples were created in Microsoft Excel™ as part of actual property evaluations; interpretations (trend lines, etc.) are derived from manual interpolation of available data.

## Production Rate as a Function of Date of Production

Usually referred to as the **Rate-Time decline curve**, the monthly production rate is plotted in a semi-log format as a function of the date of production with Bbl and/or MCF per month on the Y-axis and Production Month on the X-axis. Figure 1 shows production data from the Neely lease, a multi-well property in the Midway-Sunset field in Kern County, from 1977 through 2019. There is a considerable amount of information being transmitted in the graph. The property is part of a cyclic steaming project. Injection peaked in about 1993 and then began to be reduced in about 1998 as the oil production-steam (heat) injection efficiency declined. The data includes Produced Oil and Water, Injected Water as Steam, and Produced Gas. Also shown is the number of Producing Days which can (assuming 30-day months) be converted to Producing Wells. The year-over-year decline in oil production after 1998 is obvious, as is the decline in the other plotted factors.

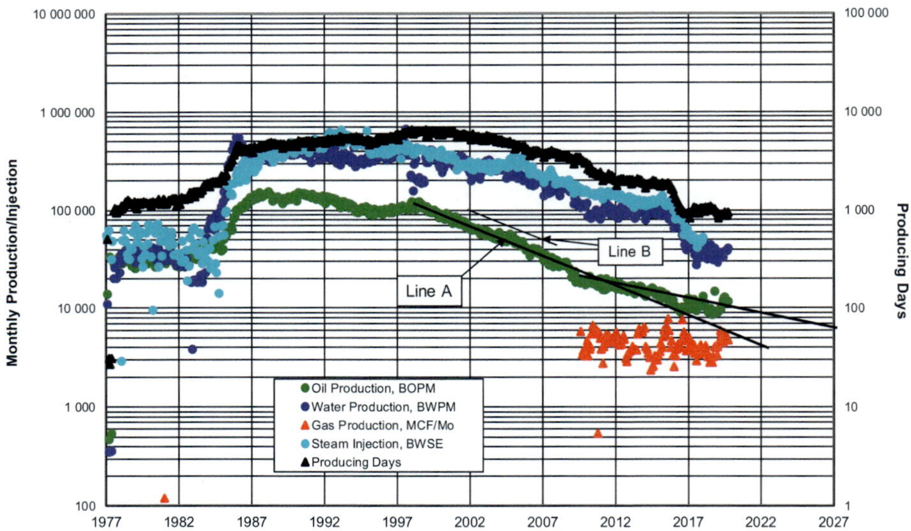


**Figure 1.** Neely Lease, Midway-Sunset field, Kern County, California

The oil production demonstrates an apparently *exponential* decline from about 1999 through 2009 of approximately 13% per year, followed by a somewhat less steep decline through 2019 of about 7% per year. The reduction in steam injection is coincident with the beginning of the decline in both oil and water. The number of producing days and, by extension, the number of producing wells, also begins to decline. The oil production decline is sufficiently established to allow extrapolation of the representative production rate into the future. The oil production decline rate can be calculated as an annual loss rate or it can be defined by use of a straight line drawn through the data points to obtain annual decline rate.

## Estimating Decline Rates

This is a good place to discuss the derivation of decline rates from graphical representations of oil & gas production. When plotted in a semi-log format, the decline rate can be defined directly from the graph using a method known as the "Laying on of the Triangles." A production decline curve such as shown in Figure 1 suggests a relatively consistent and stable decline at about 13% that could be read almost by "eyeball" if the X-axis were annual. However, the X-axis in Figure 1 is a five-year range. Let's call the 13% decline "Line A." A couple of drafting triangles (optimally a 45 and a 30-60) can be used to estimate the annual decline rate. In Figure 1, an exponential decline trend is drawn through the data. Placing the two triangles together with the 45 supporting the 30-60 and the long edge of the 30-60 aligned with Line A, slide the 30-60 upward until the long edge intersects the next higher 10s line on the Y-axis at 2002; this is the 100,000 Bbl line. Draw a line (Line B) parallel to Line A through the next X-axis line at 2007. Moving over to the 2007 intercept, the production rate is at about 74,000 Bbl. The decline is 26,000 Bbl per month over 5 years. The annual decline rate can be calculated as follows:

$$D_a = -\log\left[\frac{\log(1-D_5)}{T}\right]$$

where ($D_a$) is the annual decline, $D_5$ is the effective decline over 5 years, and T is the effective number of years represented. In the example of Figure 1:

$$D_a = -\log\left[\frac{\log(1-0.5)}{5}\right]$$, the annual decline is 0.13 or 13%

Figure 2 shows a Rate-Time production graph for the Lehman lease, a multi-well property in the Kern Front field, also in Kern County, from 1977 through 2015. The graph shows much the same information as the Midway-Sunset property but with a few differences. This property was under primary production through about 2003 when continuous steam injection was begun. The steam injection has continued through 2015. There are arguably two exponential decline trends shown. The first is the primary decline (about 5% per year) from 1977 through 2003; the second is the decline from the enhanced recovery (about 8%) from mid-2010 through 2015. Had primary production continued, extrapolation of production using an exponential decline could have been used for future production. The decline starting in 2010 could also be extrapolated although there may be a question as to whether an exponential or hyperbolic approach would be appropriate. As noted above, exponential is commonly used as a proxy for hyperbolic in situations where the hyperbolic trend nears asymptotic. Figure 2 shows a straight line through the data points describing the exponential decline for the 2010-2015 period. Future production is estimated by projecting a representative initial rate, i.e., 21,000 BOPM, declining at 8% per year, assuming that steaming continues.

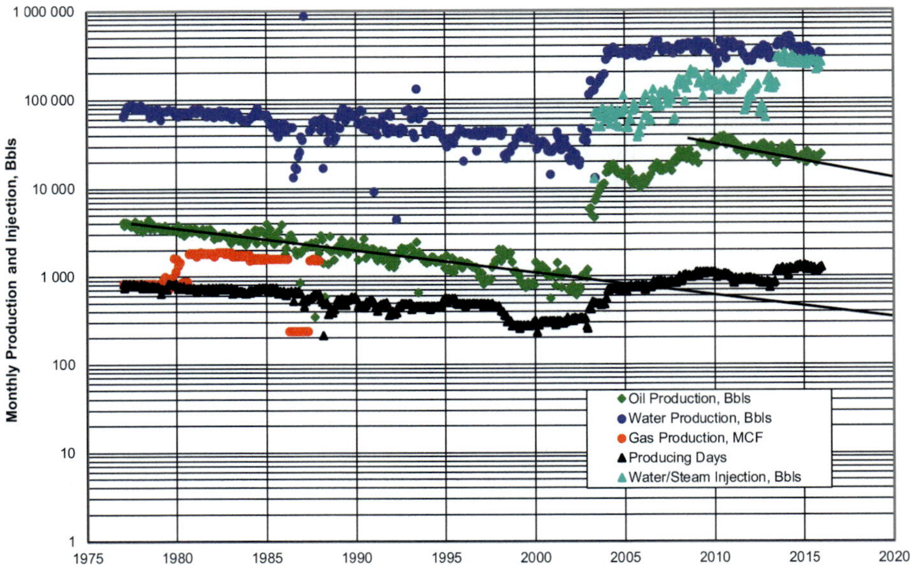


**Figure 2.** Lehman Lease, Kern Front field, Kern County, California

Figure 3 shows a Rate-Time decline curve for the Slaughter Unit 1-1, a four-well gas-producing property in Pittsburg County, Oklahoma. The decline form is clearly hyperbolic for most of the life of production but does appear to be nearing exponential after 2021. A close look at the most recent period suggests an exponential decline of about 10% per year.

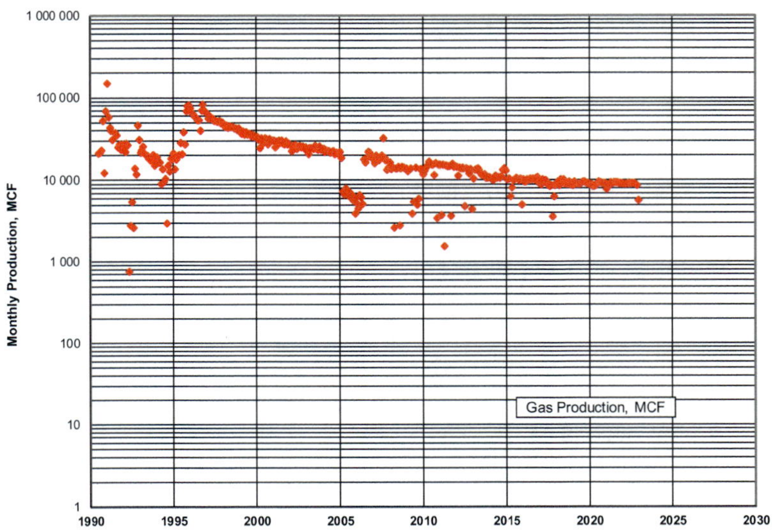


**Figure 3.** Slaughter Unit 1-1, Pittsburg County, Oklahoma

Figure 4 shows production from the Marina Lease in the South Belridge field in Kern County from 1977 through 2019. The graph demonstrates several periods of changing production (including a gap in 2018) that is not particularly easy to evaluate. However, the period from about 2000 through 2017 suggests three intervals of reasonably stable and (likely) exponential performance in 2000-2005, 2008-2012, and 2014-2017. Projection of the last period decline of about 6.5% per year could provide a foundation for a future production estimate.

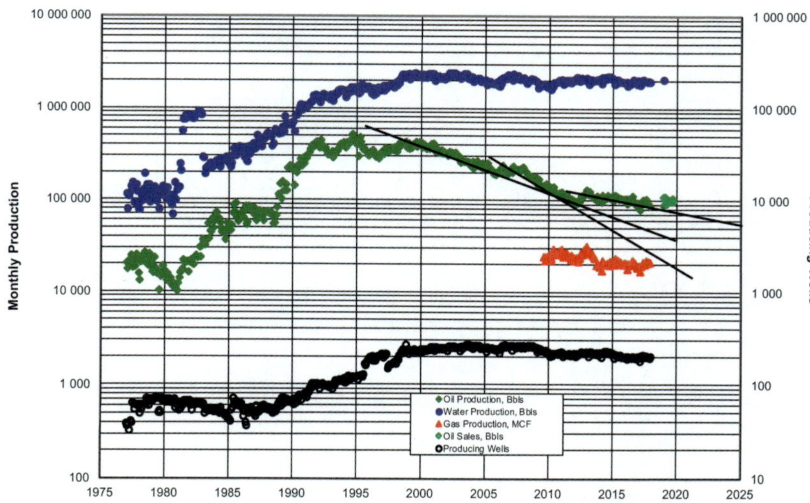


**Figure 4.** Marina lease, South Belridge field, Kern County, California

The examples shown above are relatively small properties; however, the same methods of graphical analysis can be applied to much larger properties.

The Elk Hills field in western Kern County was in production for many years as part of the Naval Petroleum Reserve. The production history shown in Figure 5 indicates that oil production increased substantially in the late 1970s when the field was opened to commercial operation. Elk Hills is a huge oil field covering roughly 74 square miles, and production is obtained from several different formations and reservoirs. It is not uncommon to evaluate the production from the different reservoirs and from separate areas of the field, but for the purpose of graphical analysis, it is useful to look at the aggregated production from the whole field. Depending on one's view of the data (which would be aided by a knowledge of field operations obtained from the operator), four exponential production periods with different decline rates could be defined as shown on the graph. The last period (2015-2022) could be used to extrapolate oil production as part of an evaluation of the Elk Hills field. The graph suggests a representative production rate of about 650,000 Bbl per month and an annual exponential decline of about

6.5%. The exponential decline could be the result of aggregating the production of multiple reservoirs. Decline curves for individual reservoirs could exhibit hyperbolic declines.

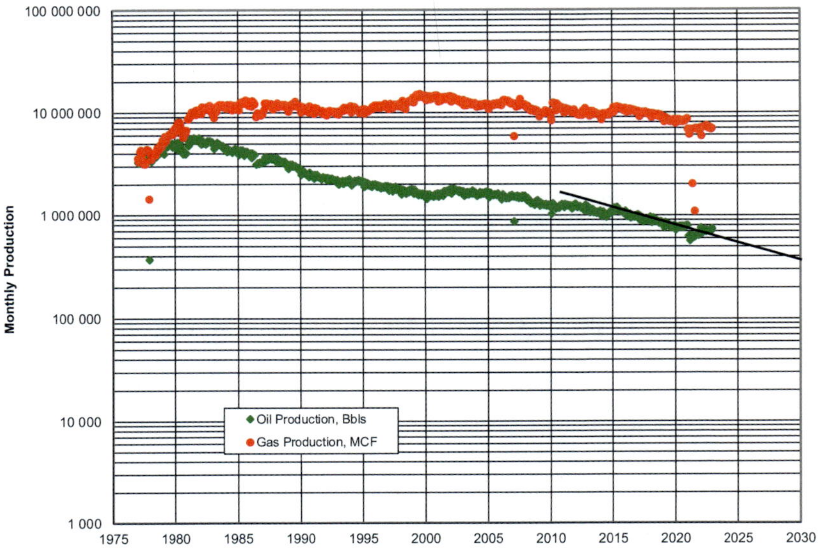


**Figure 5.** Elk Hills field, Kern County, California

Another large property which presents some interesting aspects is the Long Beach Unit (LBU) which is part of the larger Wilmington field in Los Angeles County. The production history of a portion of the life of the LBU is shown as Figure 6. Without going into a lot of detail, the Wilmington field has been in production since the early 1900s and produced so much oil and water that subsidence in Long Beach City and harbor became a serious problem. The LBU was created and developed in the late 1960s as a water injection project to attempt to arrest the subsidence. The LBU consists of several stacked reservoirs broken into over 30 fault blocks, each of which could be evaluated using Rate-Time graphical analysis.

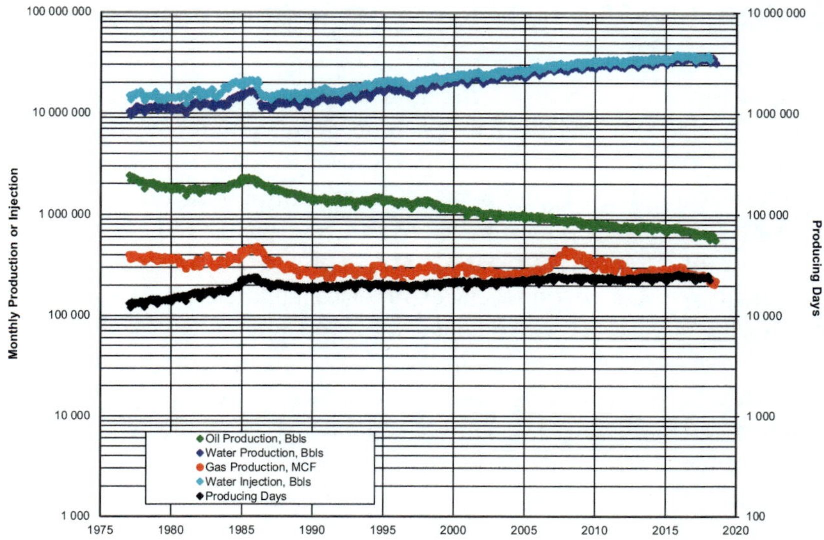


**Figure 6.** Long Beach Unit, Wilmington field, Los Angeles County, California

The graph in Figure 7 suggests an exponential decline of about 4% from 2000 through 2015, followed by a slightly steeper decline starting in 2016. Extrapolation of the latter period decline allows an estimate of future production. The correct way to evaluate Elk Hills or LBU is by separation of the projects into reservoirs and/or fault blocks to allow for the different and unique geologic and reservoir fluid characteristics found therein.

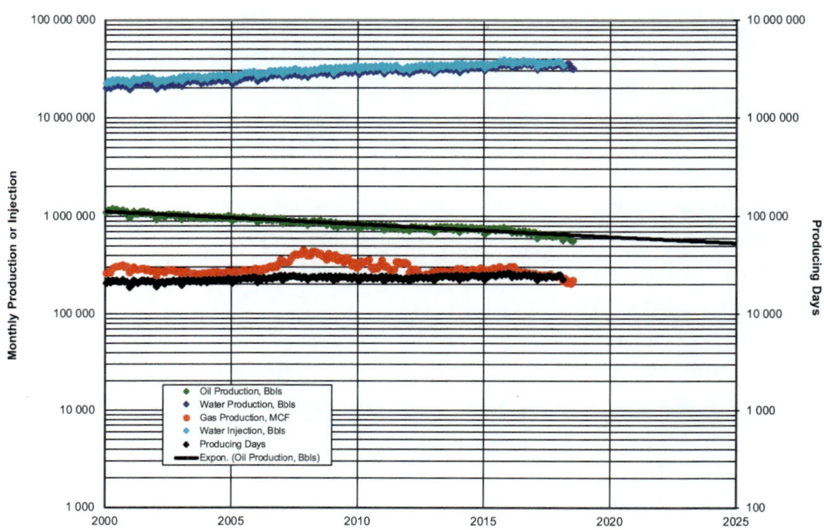


**Figure 7.** Long Beach Unit, Wilmington field, Los Angeles County, California

## Production Rate as a Function of Cumulative Production

If a property has an history of reasonably continuous operation and production, a plot of oil production rate as a function of cumulative oil production often yields a reasonably straight line. Extrapolation of this line to an economic limit production rate yields the ultimate oil or gas recovery under the assumed operating and economic conditions. Figure 8 shows monthly oil production as a function of cumulative oil production for the Midway-Sunset (Neely) property. The information from the graph is similar to the data provided by the Rate-Time graph in suggesting two decline periods. The extrapolation of the early Rate-Cumulative decline and the later, much more definitive, Rate-Cumulative curve leads to an interesting observation; the end points of the two extrapolations are very close at about 31.5 million Bbl. The explanation is that cyclic steaming accelerates primary oil recovery but does not increase ultimate recovery (see below – Primary Production Stimulation).

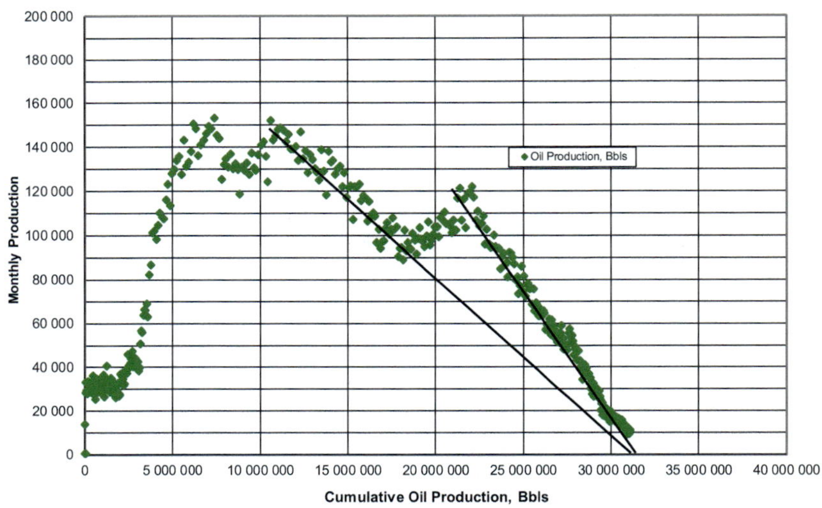


**Figure 8.** Neely Lease, Midway-Sunset field, Kern County, California

The Rate-Cumulative graph (Figure 9) for the Kern Front (Lehman) property is not as definitive as the Midway-Sunset property and is subject to interpretation, but it does suggest an end point (about 6.5 million Bbl) that could be compared to the Rate-Time extrapolation. In both examples, the end point is shown as a production rate of zero, but a rate derived from economic limit analysis would be preferable.

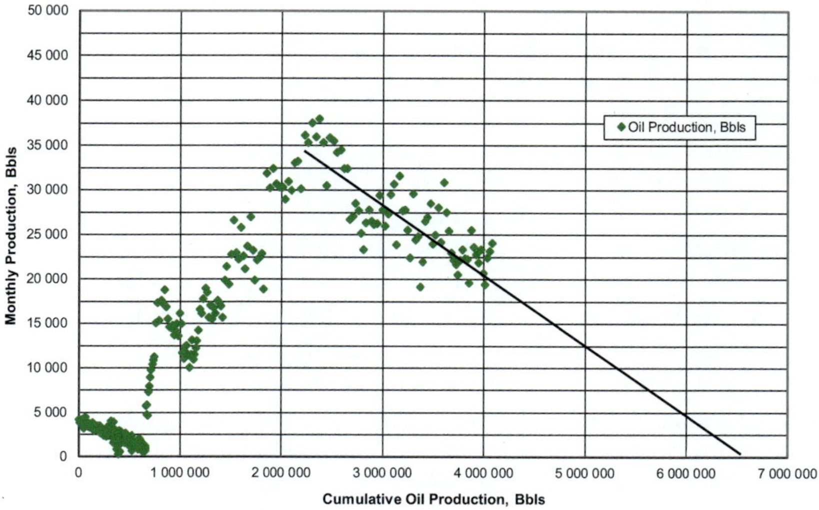


**Figure 9.** Lehman Lease, Kern Front field, Kern County, California

A Rate-Cumulative graph for the Marina lease (Figure 10) is also reasonably definitive when projected over the entire 2000 through 2017 period. However, the analysis could also consider separate projections for two or three segments of the graph, particularly the last segment starting at about 120,000 Bbl per month (~85,000,000 Bbl cumulative), to an estimated ultimate recovery of about 110 million Bbl.

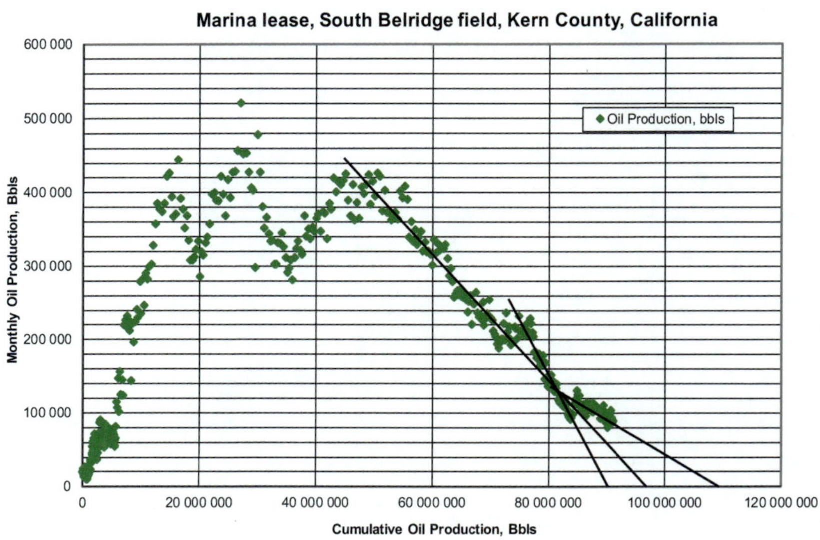


**Figure 10.** Marina lease, South Belridge field, Kern County, California

The Rate-Cumulative and the Rate-Time constructions are two parts of the estimation of future production using actual performance. This approach ties together two parts of the decline equations ($q$ and $N_p$). The Rate-Time extrapolation provides an estimate of annual production rates starting from a representative rate and future rates declining at a derived percentage over time. The Rate-Cumulative projection identifies the ultimate recovery from the property and provides a check on the Rate-Time projection. The ultimate recovery can also be compared to the sum of the decline curve extrapolated volumes. The Rate-Cumulative extrapolation method can be applied to entire fields, reservoirs, segments of reservoirs, and to individual wells (assuming closed drainage area) and is remarkably effective, particularly when there is a long, relatively uninterrupted production history.

The example shown in Figure 11 is suggested by an evaluation of the Elk Hills field where, as noted above, oil & gas production were obtained from several distinct reservoirs under differing operating systems. As part of the evaluation, a plot was constructed of the total field oil production rates relative to the accumulated oil production over the same period. It was found that the total field remaining production estimated from this approach was quite close to the total expected remaining production obtained from the detailed evaluation of the individual reservoirs. The graph suggests more than one relation of monthly production rate to cumulative production as represented by three projection relations. The points of intersection with the zero rate identify three possible ultimate recoveries which are the result of continuing development and changing operations in the field. Here again, the Rate-Time projection can be compared to the Rate-Cumulative projections to estimate annual production and ultimate recovery.

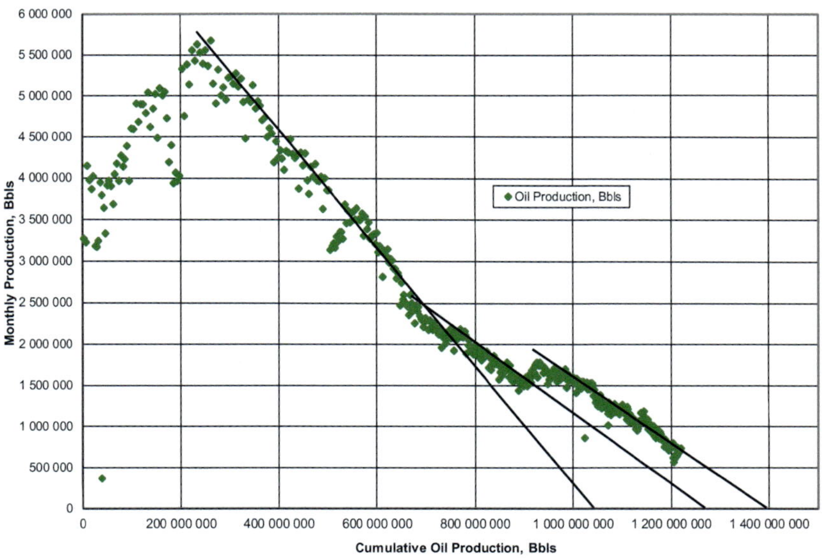


**Figure 11.** Elk Hills field, Kern County, California

A similar result is obtained from an analysis of the Long Beach Unit using a Rate-Cumulative projection (Figure 12) to control the Rate-Time projection. The intercept of the LBU production with the zero-oil production rate suggests an ultimate recovery at about 820 million Bbl.

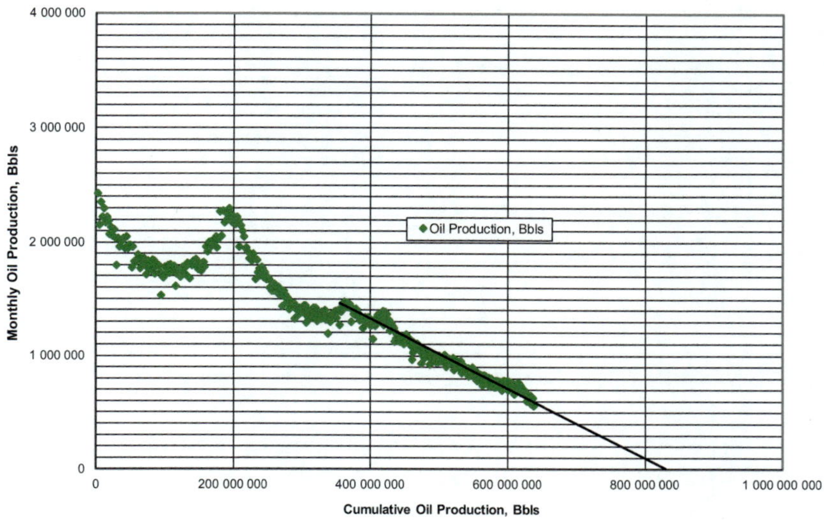


**Figure 12.** Long Beach Unit, Wilmington field, Los Angeles County, California

**Note:** The use of zero production rate as an intercept for the Rate-Time and Rate-Cumulative examples is for convenience of discussion only. In the evaluation of specific properties, the projection should terminate at the economic limit of production (see below – Economic Limit).

## Production Rate as a Function of Months Since First Production

Another form of Rate-Time decline curve is a relation of the production rate to the number of months since the start or re-start of production. It is particularly useful in the early stages of production where a large portion of ultimate recovery often occurs in the first 24-36 months from first production. The construction has been shown to be especially useful in modeling production from hydraulically fractured shale reservoirs such as the Marcellus, Barnett, and Eagle Ford shales. Figure 13 shows Barnett Shale gas production from two wells (Riley Unit 1H and 2H) in the Newark East field in Johnson County, Texas. The production history over about 92 months is clearly hyperbolic and does not yet suggest becoming exponential.

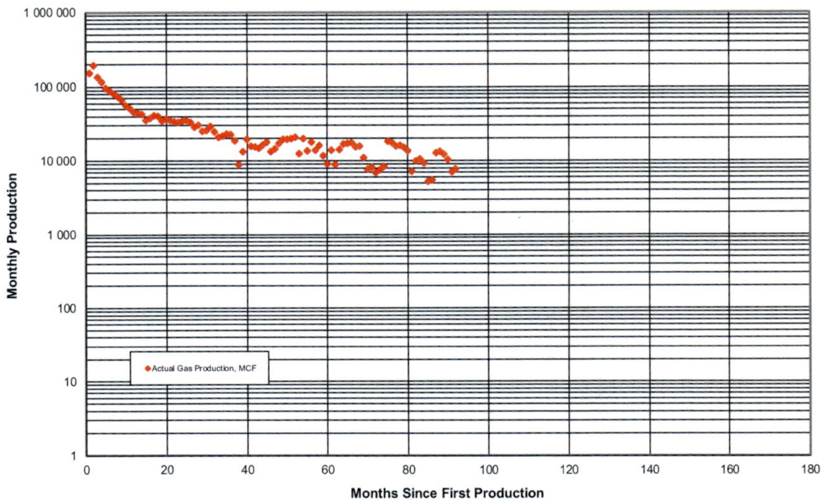


**Figure 13.** Riley Unit 1H and 2H, Newark East (Barnett Shale) field, Johnson County, Texas

Figure 14 shows the much shorter life (~18 months) of two Eagle Ford wells (Kennedy 4H and 5H) in the Eagleville field, Live Oak County, Texas. Here too, the hyperbolic decline is clearly demonstrated. This is not to say that a Rate-Time graph would not also demonstrate the hyperbolic decline form but the use of the time since first production augments the conventional Rate-Time result. As discussed below, the Months from First Production depiction form is a useful foundation for history-matching analysis.

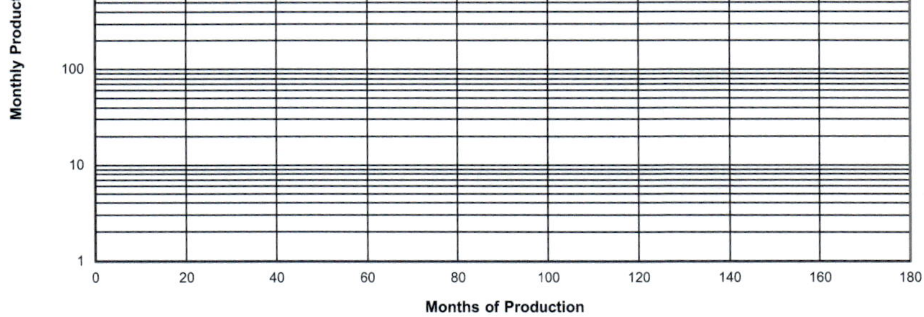


**Figure 14.** Kennedy 4H & 5H, Eagleville (Eagle Ford) field, Live Oak County, Texas

# Primary Production in Gas Reservoirs

All of the above discussion referred to primary production from oil wells. However, all graphical methods and conditions are equally applicable to gas reservoirs and properties. Gas reservoirs are governed by fluid flow in porous media under the same laws of physics as other fluids. Gas production exhibits exponential, hyperbolic, and harmonic performance that can be used to extrapolate future production rates. Some equations may have to be modified to account for the properties of natural gas rather than oil, but functionally the approaches are the same. Estimates of gas production from fractured shale reservoirs were found to follow hyperbolic decline forms and can be used for history-matching to production rate as a function of months of production decline curves. Figures 15 through 19 are Rate-Time graphs that depict primary natural gas production from five properties in different locations and from one or more wells that are horizontal, hydraulically fractured shale completions.

The Slaughter 1-1 Unit (Figure 15), as noted above regarding Figure 3, is a long-lived gas-producing property in Pittsburg County, Oklahoma. The production performance is clearly hyperbolic that could be trending toward exponential in the latter (2018-2022) phase of production.

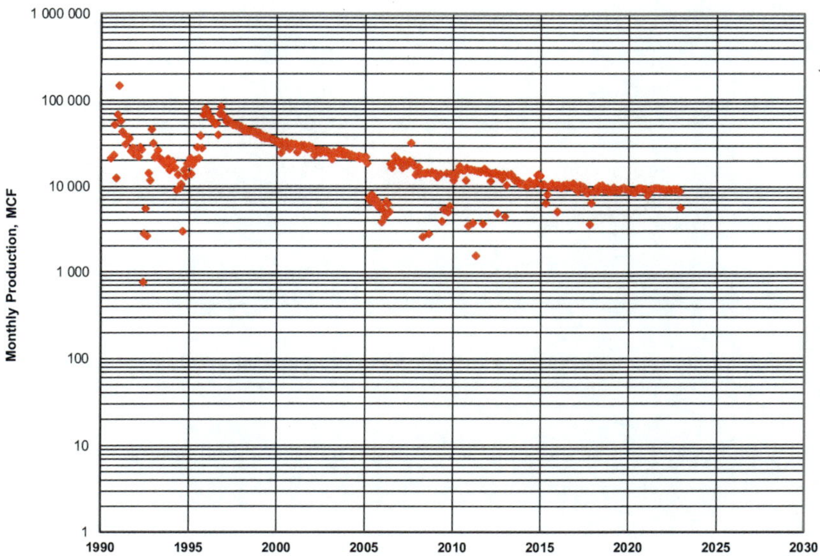


**Figure 15.** Slaughter Unit 1-1, Pittsburg County, Oklahoma

The Gamble 2 (Figure 16) is a single well located in the Bethany Longstreet field, DeSoto Parish, Louisiana, producing from the Haynesville shale. The production decline is hyperbolic though perhaps not as clearly defined as the Slaughter property. There is some tendency toward exponential in later years.

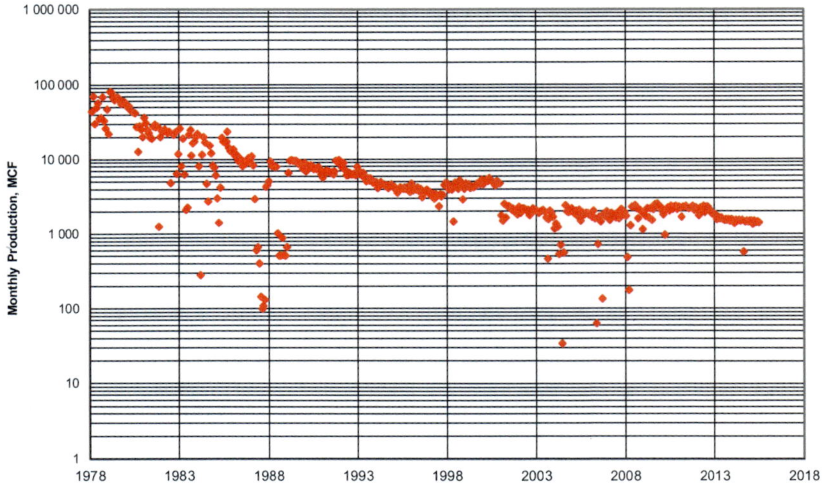


**Figure 16.** Gamble 2, Bethany Longstreet field, DeSoto Parish, Louisiana

The Bobst Unit (Figure 17) is a four-well property producing from the Marcellus Shale in Lycoming County, Pennsylvania. Production to year-end 2022 is hyperbolic trending toward exponential after 50 months.

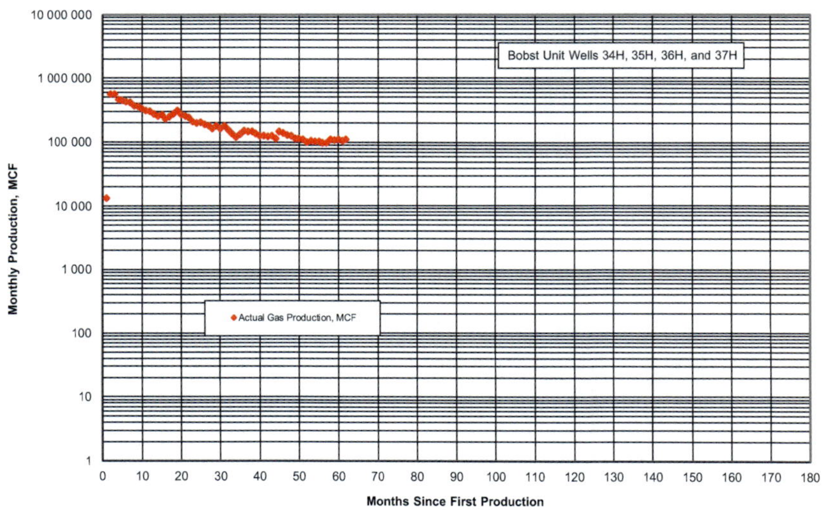


**Figure 17.** Bobst Unit, Cogan House Township, Lycoming County, Pennsylvania

The White Gas Unit (Figure 18) is a multi-well property producing from the Minden (Cotton Valley) formation in Rusk County, Texas. It is clearly hyperbolic, arguably becoming linear in the later life.

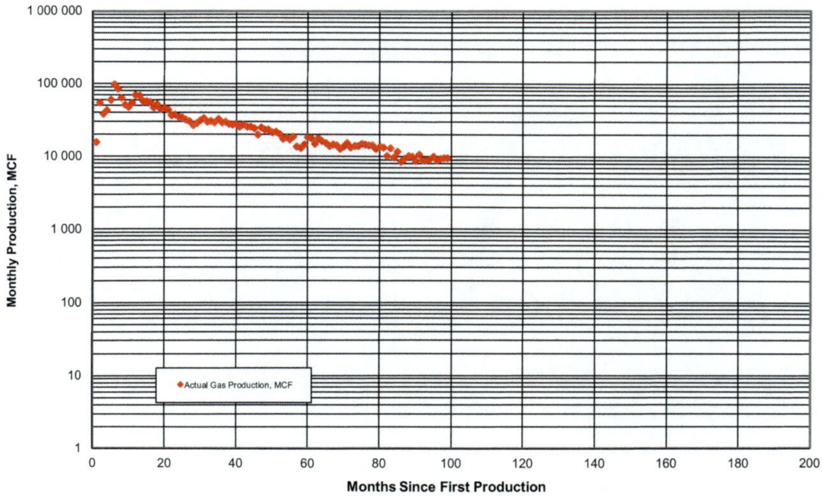


**Figure 18.** White Gas Unit, Minden (Cotton Valley) field, Rusk County, Texas

The Stewart Title 7-9 1-19H (Figure 19) is a single well producing from the "B-43" Fayetteville Shale field in White County, Arkansas, where gas production demonstrates a classic hyperbolic decline that has not yet begun to approach asymptotic.

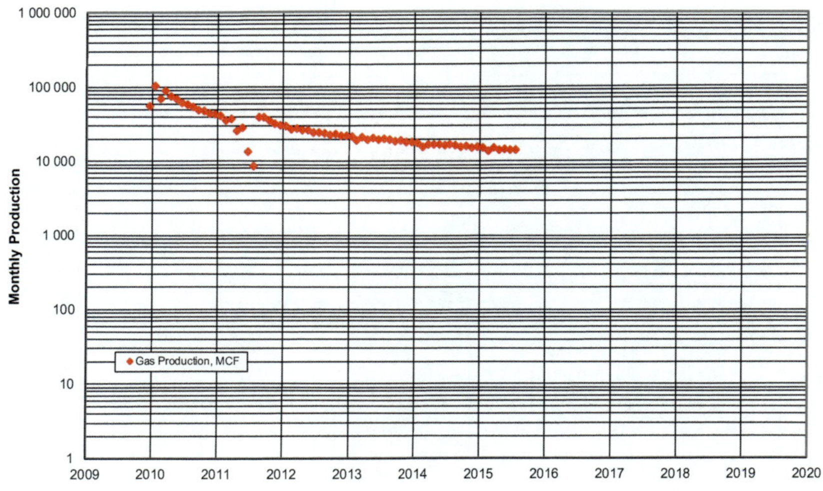


**Figure 19.** Stewart Trust 7-9 1-19H, B-43 (Fayetteville) field, White County, Arkansas

As should be expected, the Rate-Cumulative model also applies to analysis of gas wells. Figures 20 and 21 show Rate-Cumulative analysis for two properties producing from the Marcellus Shale in north-central Pennsylvania. In this case the Rate-Cumulative plots as a semi-log function.

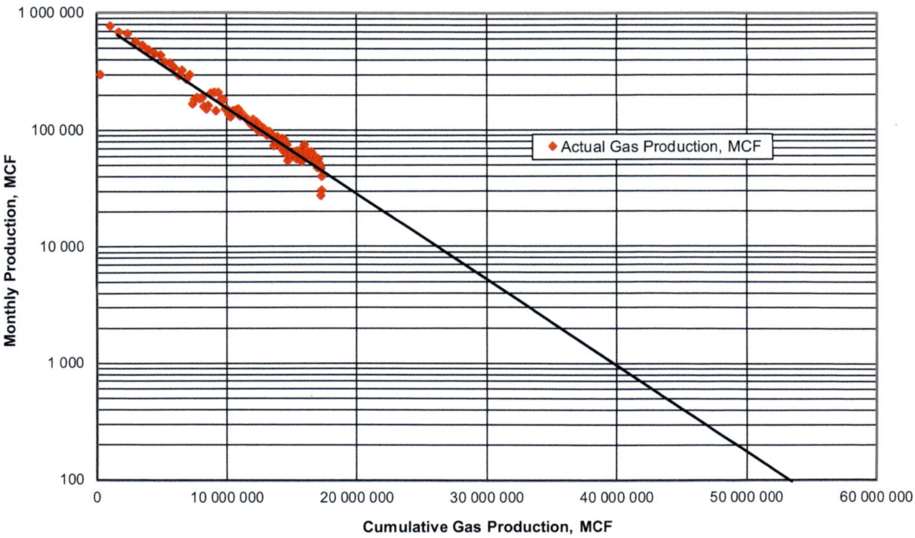


**Figure 20.** Bobst Mountain Hunting Club Lease, Lycoming County, Pennsylvania

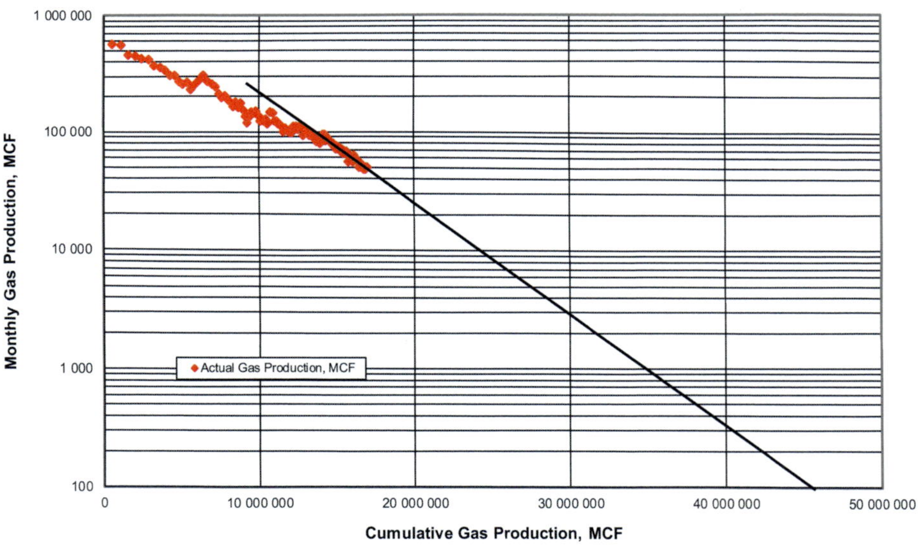


**Figure 21.** Bobst Unit, Cogan House Township, Lycoming County, Pennsylvania

## Water-Oil-Ratio (WOR) as a Function of Cumulative Oil Production

In water-drive reservoirs and, to some extent, in many water-flood or steam-flood operations, ultimate recovery can be estimated by plotting the ratio of water production to oil production (Water-Oil Ratio or WOR, Bbl/Bbl) as a function of cumulative oil recovery. Production and disposal of wastewater is a major component of operating costs and can be a factor in defining the economic limit (see below – Economic Limit) for a property. The maximum WOR that can be sustained under the prevailing economic conditions can be used as a limit of production. The cumulative production at that point is ultimate recovery and remaining recovery can be estimated as the difference between ultimate and cumulative recovery to date. A variation on this method is to use water-cut (water as a percentage of total produced fluid) rather than WOR. Figure 22 illustrates this approach for the Long Beach Unit where WOR is approaching 60 BW/BO and water-cut is nearing 100%. Depending on the economic analysis, the WOR (or Water Cut) as a function of cumulative oil production approach could be used to define ultimate recovery between 600 million and 700 million Bbl. In comparison, the Rate-Cumulative for the LBU projection to a zero-oil production rate suggested 820 million Bbl of ultimate recovery.

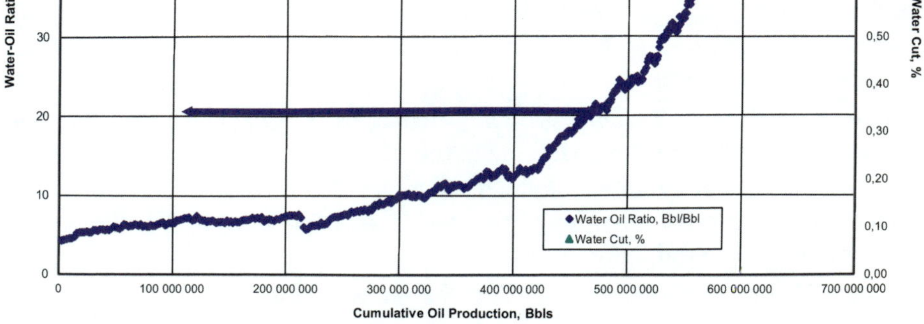


**Figure 22.** Long Beach Unit, Wilmington field, Los Angeles County, California

## Gas-Oil Ratio (GOR) as a Function of Cumulative Oil Production

Particularly in solution-gas drive reservoirs, a plot of cumulative oil as a function of the ratio of produced gas to produced oil (GOR, SCF/Bbl) can be used to estimate ultimate oil recovery at a limiting gas production where GOR is a proxy for gas production, and to estimate periodic and ultimate GOR which can then be converted to associated gas production. Conversely, if gas volume is a limitation on production (such as where gas cannot be sold, flared, or otherwise disposed), then GOR as a function of cumulative oil production can be used to estimate ultimate recovery.

## History-Matching

While extrapolation of Rate-Time and Rate-Cumulative decline curves can be useful and provide a reasonable estimate of future production and ultimate recovery, the results may not always be conclusive when the property is in the early stages of production. A history-matching approach can be useful, particularly where early production demonstrates a hyperbolic decline form. In this procedure, the actual production is rendered graphically as barrels or MCF per unit time in months since the start of production. The hyperbolic equation is used to calculate production rates over the same period as the actual production. A computer-aided trial and error process is then used to attempt to calculate a match for the historical production. The calculated rates are overlaid on the actual production graph until a reasonable correlation is obtained. The calculation and graph can then be extrapolated to obtain future production rates and cumulative production. Needless to say, the more actual production history that is available, the better the match will be. History-matching requires a certain amount of work, but the results can be rewarding. The rapid early decline in fractured shales is clearly demonstrated in a rate-month plot. Further, the ($D_i$) and $n$ values derived as part of the history-match calculations can be considered for application to evaluation of similar wells.

Figures 23 through 28 show production analysis of wells or small groups of wells evaluated using history-matching in the Marcellus Shale, Haynesville Shale, Fayetteville Shale, Barnett Shale, and Eagle Ford formation. In each case the actual production was plotted as volume per month as a function of the number of Months Since First Production.

Figure 23 is a group of four Marcellus wells in Lycoming County, Pennsylvania. The wells have been in production for about 63 months at the time of evaluation. The decline is obviously hyperbolic. The history-matching production is calculated by a trial-and-error process of adjusting the initial decline rate ($D_i$) and ($n$) values to obtain a production projection and then overlaying that projection on the actual history (Exhibit 2). This may have to be done multiple times until a match, as shown on the graph, is found.

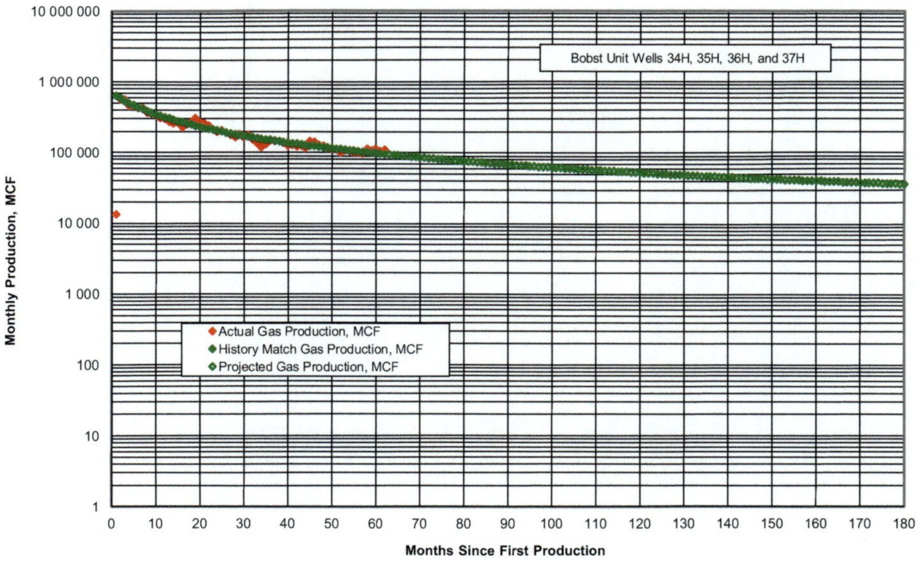


**Figure 23.** Bobst Unit, Cogan House Township, Lycoming County, Pennsylvania

Also as shown, the history-match can then be projected to estimate future production. This approach can take time, but thanks to Intel and Microsoft, the result is a workable model. The same procedure is used to obtain history-match production estimates for the properties in Figures 24 through 28.

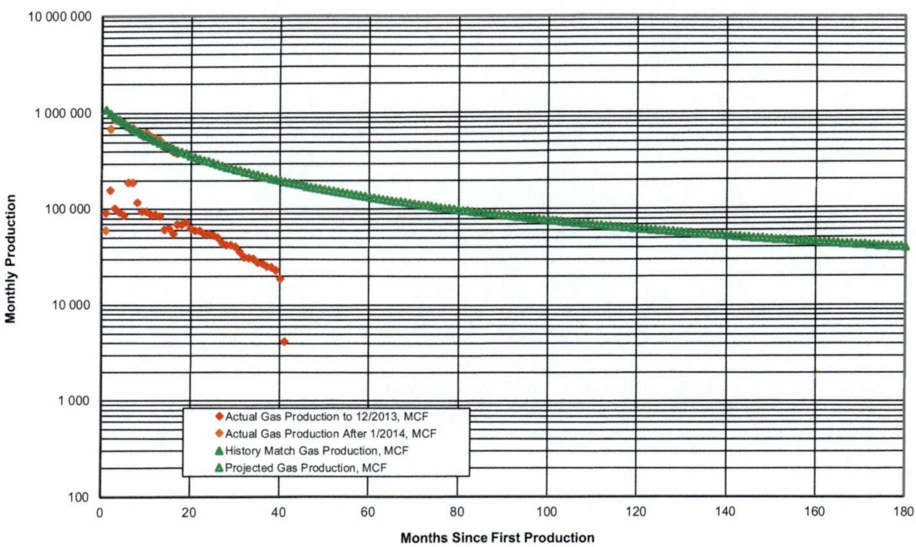


**Figure 24.** Sample 4-3H, Thorn Lake (Haynesville) field, Red River Parish, Louisiana

Initial Decline, Di = 0.125

| Production Months | Gas Production MCF | Time t | n 1.5 | Gas Production MCF qi 180000 | Production Months | Gas Production MCF | Time t | n 1.5 | Gas Production MCF qi 180000 |
|---|---|---|---|---|---|---|---|---|---|
| 1 | 5,542 | 1 | 0.891752 | 160,515 | 91 | 24,978 | 91 | 0.145261 | 26,147 |
| 2 | 66,831 | 2 | 0.808720 | 145,570 | 92 | 25,036 | 92 | 0.144264 | 25,968 |
| 3 | 131,585 | 3 | 0.742654 | 133,678 | 93 | 25,395 | 93 | 0.143284 | 25,791 |
| 4 | 121,666 | 4 | 0.688612 | 123,950 | 94 | 23,982 | 94 | 0.142321 | 25,618 |
| 5 | 118,545 | 5 | 0.643436 | 115,819 | 95 | 25,331 | 95 | 0.141374 | 25,447 |
| 6 | 105,862 | 6 | 0.605007 | 108,901 | 96 | 26,282 | 96 | 0.140442 | 25,280 |
| 7 | 101,887 | 7 | 0.571846 | 102,932 | 97 | 25,748 | 97 | 0.139526 | 25,115 |
| 8 | 96,201 | 8 | 0.542884 | 97,719 | 98 | 23,154 | 98 | 0.138624 | 24,952 |
| 9 | 87,687 | 9 | 0.517330 | 93,119 | 99 | 24,644 | 99 | 0.137737 | 24,793 |
| 10 | 86,979 | 10 | 0.494586 | 89,025 | 100 | 22,054 | 100 | 0.136864 | 24,635 |
| 11 | 80,992 | 11 | 0.474187 | 85,354 | 101 | 23,534 | 101 | 0.136004 | 24,481 |
| 12 | 78,751 | 12 | 0.455769 | 82,038 | 102 | 23,145 | 102 | 0.135158 | 24,328 |
| 13 | 76,893 | 13 | 0.439041 | 79,027 | 103 | 23,803 | 103 | 0.134325 | 24,179 |
| 14 | 65,819 | 14 | 0.423768 | 76,278 | 104 | 25,884 | 104 | 0.133505 | 24,031 |
| 15 | 73,841 | 15 | 0.409757 | 73,756 | 105 | 20,000 | 105 | 0.132697 | 23,885 |
| 16 | 58,767 | 16 | 0.396850 | 71,433 | 106 | 22,184 | 106 | 0.131901 | 23,742 |
| 17 | 66,567 | 17 | 0.384914 | 69,284 | 107 | 23,206 | 107 | 0.131117 | 23,601 |
| 18 | 61,123 | 18 | 0.373836 | 67,291 | 108 | 23,326 | 108 | 0.130345 | 23,462 |
| 19 | 60,944 | 19 | 0.363523 | 65,434 | 109 | 23,793 | 109 | 0.129584 | 23,325 |
| 20 | 60,072 | 20 | 0.353892 | 63,701 | 110 | 21,765 | 110 | 0.128834 | 23,190 |
| 21 | 53,544 | 21 | 0.344875 | 62,078 | 111 | 23,126 | 111 | 0.128094 | 23,057 |
| 22 | 48,961 | 22 | 0.336411 | 60,554 | 112 | 22,302 | 112 | 0.127365 | 22,926 |
| 23 | 55,431 | 23 | 0.328449 | 59,121 | 113 | 22,377 | 113 | 0.126647 | 22,796 |
| 24 | 52,172 | 24 | 0.320941 | 57,769 | 114 | 24,066 | 114 | 0.125938 | 22,669 |
| 25 | 49,789 | 25 | 0.313848 | 56,493 | 115 | 20,964 | 115 | 0.125240 | 22,543 |
| 26 | 47,623 | 26 | 0.307134 | 55,284 | 116 | 20,176 | 116 | 0.124551 | 22,419 |
| 27 | 52,154 | 27 | 0.300768 | 54,138 | 117 | 18,164 | 117 | 0.123871 | 22,297 |
| 28 | 46,994 | 28 | 0.294723 | 53,050 | 118 | 19,832 | 118 | 0.123200 | 22,176 |
| 29 | 48,623 | 29 | 0.288972 | 52,015 | 119 | 20,093 | 119 | 0.122539 | 22,057 |
| 30 | 45,172 | 30 | 0.283493 | 51,029 | 120 | 22,571 | 120 | 0.121886 | 21,940 |
| 31 | 43,242 | 31 | 0.278267 | 50,088 | 121 | | 121 | -0.121242 | -21,824 |
| 32 | 41,069 | 32 | 0.273276 | 49,190 | 122 | | 122 | -0.120607 | -21,709 |
| 33 | 40,578 | 33 | 0.268502 | 48,330 | 123 | | 123 | -0.119979 | -21,596 |
| 34 | 41,759 | 34 | 0.263932 | 47,508 | 124 | | 124 | -0.119360 | -21,485 |
| 35 | 40,645 | 35 | 0.259551 | 46,719 | 125 | | 125 | -0.118749 | -21,375 |
| 36 | 41,622 | 36 | 0.255348 | 45,963 | 126 | | 126 | -0.118145 | -21,266 |
| 37 | 42,632 | 37 | 0.251311 | 45,236 | 127 | | 127 | -0.117549 | -21,159 |
| 38 | 37,899 | 38 | 0.247429 | 44,537 | 128 | | 128 | -0.116961 | -21,053 |
| 39 | 40,464 | 39 | 0.243694 | 43,865 | 129 | | 129 | -0.116380 | -20,948 |
| 40 | 36,829 | 40 | 0.240097 | 43,218 | 130 | | 130 | -0.115806 | -20,845 |
| 41 | 36,646 | 41 | 0.236630 | 42,593 | 131 | | 131 | -0.115239 | -20,743 |
| 42 | 36,275 | 42 | 0.233286 | 41,991 | 132 | | 132 | -0.114678 | -20,642 |
| 43 | 42,367 | 43 | 0.230057 | 41,410 | 133 | | 133 | -0.114125 | -20,543 |
| 44 | 40,214 | 44 | 0.226937 | 40,849 | 134 | | 134 | -0.113578 | -20,444 |
| 45 | 40,345 | 45 | 0.223921 | 40,306 | 135 | | 135 | -0.113038 | -20,347 |
| 46 | 41,235 | 46 | 0.221004 | 39,781 | 136 | | 136 | -0.112504 | -20,251 |
| 47 | 41,632 | 47 | 0.218179 | 39,272 | 137 | | 137 | -0.111977 | -20,156 |
| 48 | 39,387 | 48 | 0.215443 | 38,780 | 138 | | 138 | -0.111455 | -20,062 |
| 49 | 40,482 | 49 | 0.212792 | 38,303 | 139 | | 139 | -0.110940 | -19,969 |
| 50 | 34,897 | 50 | 0.210220 | 37,840 | 140 | | 140 | -0.110430 | -19,877 |
| 51 | 37,195 | 51 | 0.207725 | 37,391 | 141 | | 141 | -0.109927 | -19,787 |
| 52 | 36,096 | 52 | 0.205303 | 36,954 | 142 | | 142 | -0.109429 | -19,697 |
| 53 | 35,058 | 53 | 0.202949 | 36,531 | 143 | | 143 | -0.108936 | -19,609 |
| 54 | 33,630 | 54 | 0.200663 | 36,119 | 144 | | 144 | -0.108450 | -19,521 |
| 55 | 33,485 | 55 | 0.198439 | 35,719 | 145 | | 145 | -0.107968 | -19,434 |
| 56 | 34,370 | 56 | 0.196276 | 35,330 | 146 | | 146 | -0.107492 | -19,349 |
| 57 | 33,400 | 57 | 0.194172 | 34,951 | 147 | | 147 | -0.107021 | -19,264 |
| 58 | 34,731 | 58 | 0.192122 | 34,582 | 148 | | 148 | -0.106555 | -19,180 |
| 59 | 32,901 | 59 | 0.190126 | 34,223 | 149 | | 149 | -0.106094 | -19,097 |
| 60 | 33,645 | 60 | 0.188181 | 33,873 | 150 | | 150 | -0.105639 | -19,015 |
| 61 | 32,897 | 61 | 0.186285 | 33,531 | 151 | | 151 | -0.105188 | -18,934 |
| 62 | 31,665 | 62 | 0.184436 | 33,198 | 152 | | 152 | -0.104741 | -18,853 |
| 63 | 34,186 | 63 | 0.182632 | 32,874 | 153 | | 153 | -0.104300 | -18,774 |
| 64 | 32,422 | 64 | 0.180872 | 32,557 | 154 | | 154 | -0.103863 | -18,695 |
| 65 | 33,967 | 65 | 0.179153 | 32,248 | 155 | | 155 | -0.103431 | -18,618 |
| 66 | 31,830 | 66 | 0.177475 | 31,946 | 156 | | 156 | -0.103003 | -18,541 |
| 67 | 33,307 | 67 | 0.175836 | 31,650 | 157 | | 157 | -0.102580 | -18,464 |
| 68 | 34,081 | 68 | 0.174233 | 31,362 | 158 | | 158 | -0.102160 | -18,389 |
| 69 | 30,496 | 69 | 0.172667 | 31,080 | 159 | | 159 | -0.101746 | -18,314 |
| 70 | 31,236 | 70 | 0.171136 | 30,804 | 160 | | 160 | -0.101335 | -18,240 |
| 71 | 31,964 | 71 | 0.169638 | 30,535 | 161 | | 161 | -0.100928 | -18,167 |
| 72 | 34,898 | 72 | 0.168172 | 30,271 | 162 | | 162 | -0.100526 | -18,095 |
| 73 | 34,303 | 73 | 0.166738 | 30,013 | 163 | | 163 | -0.100127 | -18,023 |
| 74 | 30,388 | 74 | 0.165334 | 29,760 | 164 | | 164 | -0.099733 | -17,952 |
| 75 | 33,078 | 75 | 0.163959 | 29,513 | 165 | | 165 | -0.099342 | -17,882 |
| 76 | 32,116 | 76 | 0.162612 | 29,270 | 166 | | 166 | -0.098955 | -17,812 |
| 77 | 31,792 | 77 | 0.161293 | 29,033 | 167 | | 167 | -0.098572 | -17,743 |
| 78 | 29,569 | 78 | 0.160000 | 28,800 | 168 | | 168 | -0.098192 | -17,675 |
| 79 | 29,826 | 79 | 0.158733 | 28,572 | 169 | | 169 | -0.097817 | -17,607 |
| 80 | 30,271 | 80 | 0.157490 | 28,348 | 170 | | 170 | -0.097444 | -17,540 |
| 81 | 28,404 | 81 | 0.156272 | 28,129 | 171 | | 171 | -0.097075 | -17,474 |
| 82 | 29,344 | 82 | 0.155076 | 27,914 | 172 | | 172 | -0.096710 | -17,408 |
| 83 | 28,551 | 83 | 0.153904 | 27,703 | 173 | | 173 | -0.096348 | -17,343 |
| 84 | 29,309 | 84 | 0.152753 | 27,496 | 174 | | 174 | -0.095990 | -17,278 |
| 85 | 27,982 | 85 | 0.151624 | 27,292 | 175 | | 175 | -0.095635 | -17,214 |
| 86 | 25,714 | 86 | 0.150515 | 27,093 | 176 | | 176 | -0.095283 | -17,151 |
| 87 | 27,160 | 87 | 0.149426 | 26,897 | 177 | | 177 | -0.094934 | -17,088 |
| 88 | 26,128 | 88 | 0.148357 | 26,704 | 178 | | 178 | -0.094588 | -17,026 |
| 89 | 26,634 | 89 | 0.147307 | 26,515 | 179 | | 179 | -0.094246 | -16,964 |
| 90 | 25,269 | 90 | 0.146275 | 26,329 | 180 | | 180 | -0.093907 | -16,903 |

**Exhibit 2.** Lycoming County, Pennsylvania producing wells

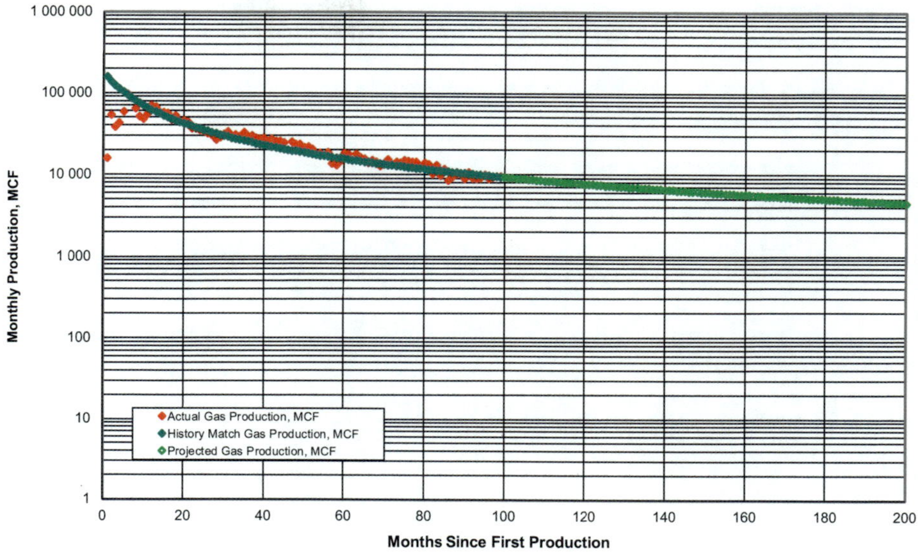


**Figure 25.** White Gas Unit, Minden (Cotton Valley) field, Rusk County, Texas

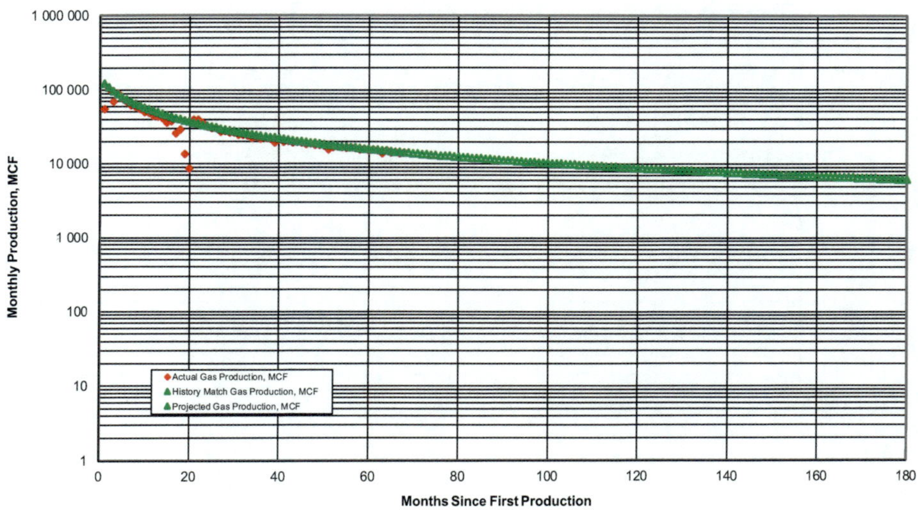


**Figure 26.** Stewart Trust 7-9 1-19H, B-43 (Fayetteville) field, White County, Arkansas

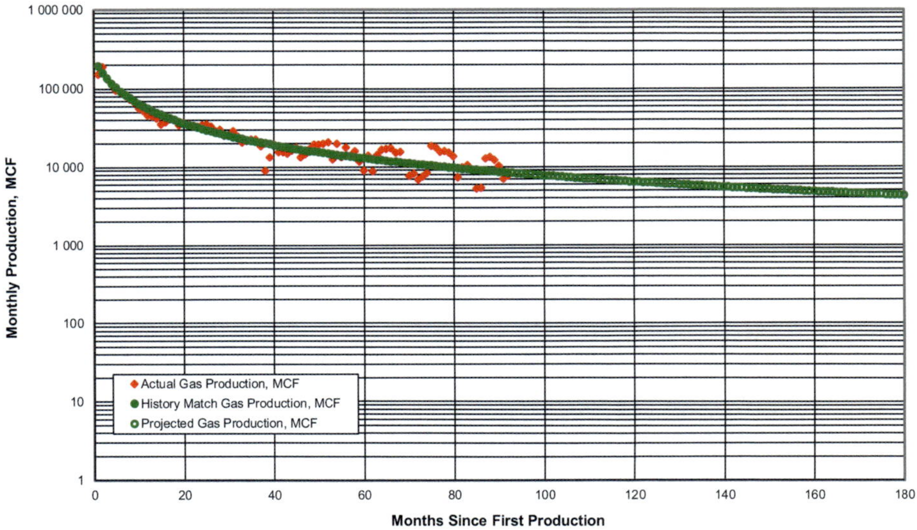


**Figure 27.** Riley Unit 1H and 2H, Newark East (Barnett Shale) field, Johnson County, Texas

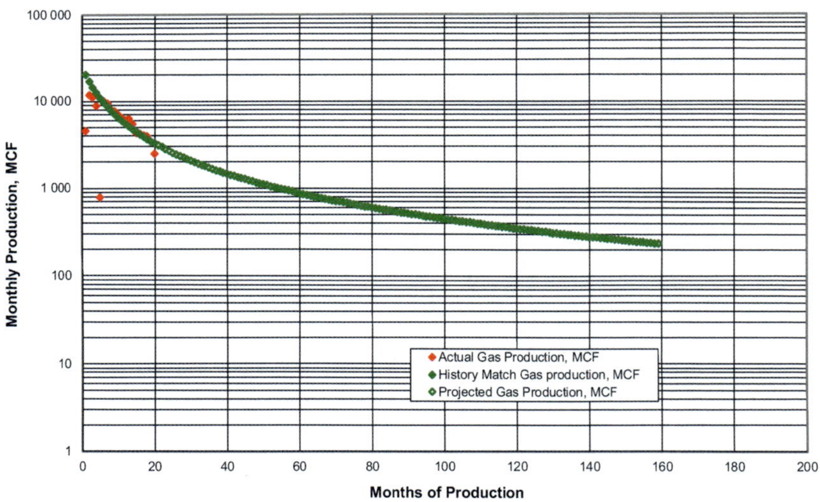


**Figure 28.** Colvin Unit B1, Eagleville field, Live Oak County, Texas

The utility of the history match method is shown by comparison of the Marcellus Shale production analysis for the nine wells in the Bobst Mountain Hunting Club property located in Lycoming County, Pennsylvania. In Figure 29 the history-match analysis for the first 63 months of production for two groups of wells found hyperbolic decline matches for both Group 1 and Group 2. There was sufficient history to allow the match of actual production to be projected as a schedule of future production rates.

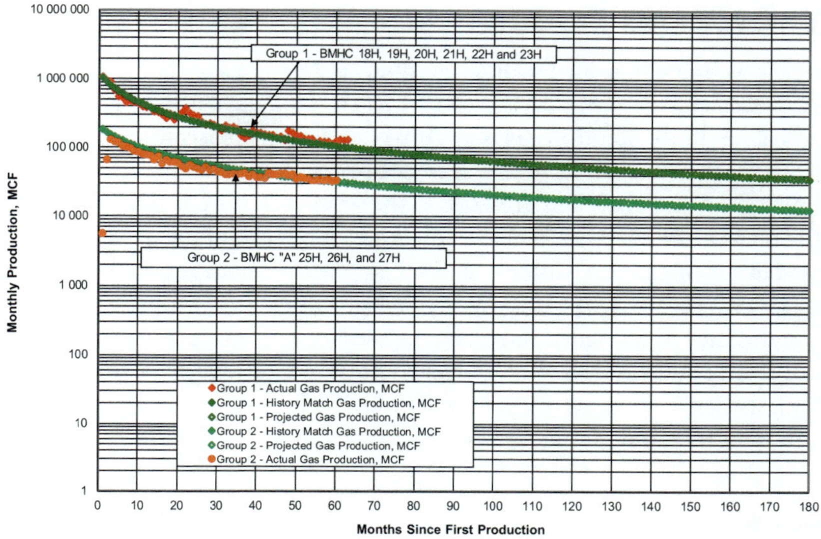


**Figure 29.** Bobst Mountain Hunting Club Lease, Lycoming County, Pennsylvania

That analysis was confirmed by a second evaluation (Figure 30) done after about 123 months of production which is shown to continue to follow the 63-month history-match projection. The analysis of Group 2 confirmed the decision to match the production for month 42 through 63 as the prevailing decline trend. This approach has been used on literally hundreds of oil & gas properties.

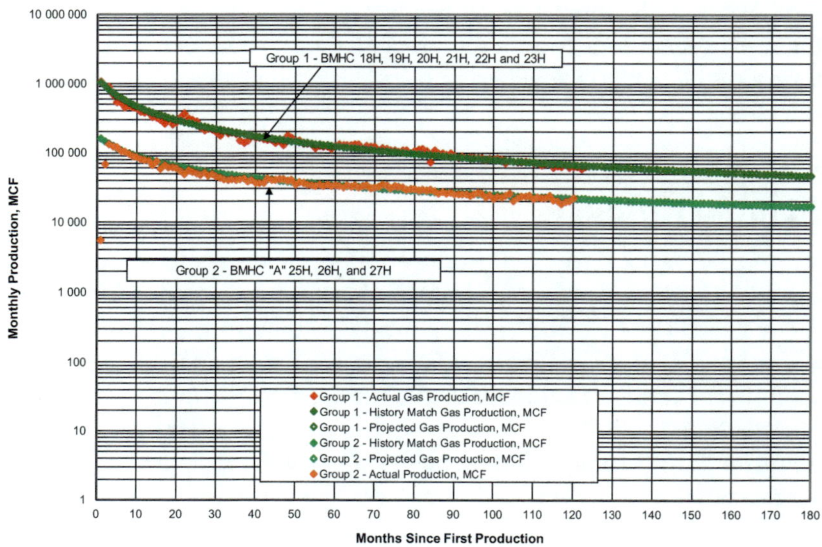


**Figure 30.** Bobst Mountain Hunting Club Lease, Lycoming County, Pennsylvania

## Associated Gas Production

Oil reservoirs commonly produce gas along with the oil, either as the result of gas being dissolved in the oil in the reservoir (solution gas), as free gas contained in a gas-cap, or both. Solution gas is a primary source of energy in many reservoirs due to the expansion of gas as reservoir pressure declines. The expanding gas forces oil to move from areas of high pressure to the wellbore. The gas is produced along with the oil as associated gas. The accepted measurement of the gas is the ratio of the gas volume in Standard Cubic Feet (SCF) to the oil volume in barrels (SCF/Bbl) known as the Gas-Oil Ratio or GOR. As oil production continues and reservoir pressure declines, GOR tends to increase. A useful relation is the change in GOR with cumulative oil production. A projection of gas production and cumulative oil production can be used to extrapolate GOR; future gas production is then estimated as a function of ongoing oil production.

## Water Production

Oil reservoirs commonly produce water along with the oil. Water drive is a primary source of energy in many reservoirs due to the encroachment of edge or bottom water; as water in the reservoir expands, pressure declines. The expanding water forces oil to move from areas of high pressure to the wellbore. The expanded water is produced along with the oil. The accepted measurement of the water produced is the ratio of the water volume in barrels to the oil volume in barrels (BW/BO) known as the Water-Oil Ratio or WOR. As oil production continues and reservoir pressure declines, WOR tends to increase. A useful relation is the change in WOR with cumulative oil production (see LBU above, Figure 22). A projection of water production and cumulative oil production can be used to extrapolate WOR; water production in barrels is then estimated as a function of oil production.

## Associated Products

Natural gas, particularly gas produced along with crude oil, can contain liquid products. These products are of variable composition depending on the source reservoir. They are known as Natural Gas Liquids or, in some contexts, as condensate. These terms have differing meanings, and application of the terms within the industry is inconsistent. For the purpose of discussion in this text, the term Natural Gas will mean Methane. Natural Gas Liquids (NGL) will include ethane, propane, butane, natural gasoline, and virtually any other hydrocarbon that can be extracted from "wet" gas. The term condensate is not used.

When actual or projected performance of natural gas production suggests that additional products may be efficiently obtained along with the produced gas, a method of estimating the volumes of those products will need to be derived, possibly from similar properties and/or reservoirs, and applied to the production projection. The term **associated products** is sometimes applied to certain hydrocarbon components of the gas such as ethane, propane, butane, and occasionally other, heavier hydrocarbons; produced gas that contains a percentage of such products is often referred to as "wet gas."

Depending on the volume of associated products and the potential market for those products, it may be economically useful to separate the entrained products. Some separation can occur in standard processing facilities simply as the result of the drop in pressure as the gas reaches the surface. Some entrained products drop out of the gas as liquids, commonly known as "natural gasoline," which (historically) was used as fuel for pumping units. Other surface facilities, often known as "chillers," can be used to entice heavier hydrocarbons to drop out of the gas. "Liquids" that are obtained this way can be used as fuel or can be mixed into produced oil to modify the gravity and viscosity of the oil.

These options are useful and can have some economic benefit but, again, depending on the potential volume and marketability of the associated products, processing through a specialized facility could have more economic value. Such facilities, usually termed "gas plants" can break out and collect the associated products from the wet gas leaving a residue of dry gas which can be sold or used as fuel. The difference between the volume of wet gas produced at the wellhead and the dry gas produced at the gas plant is termed **shrinkage**. Where wet gas is processed this way, the volume and value of each product, whether sold or used on the property, can be factored into the evaluation of the property as either additional income or as an offset to fuel costs. Sales of dry gas and/or individual products such as ethane, butane, propane, etc., comprise potential income streams that add to the value of the property.

## Family Curves

Occasionally evaluation projects involve data from multiple wells and/or from small well groups in different parts of a field or from similar properties or wells with similar operating/reservoir characteristics where evaluation of the individual wells does not provide a satisfactory result. The use of **family curves** may be efficient. In this approach, wells may be grouped based on a more or less common criterion (location, source reservoir, date of production, completion interval, and/or method of completion) in an attempt to extract a projection that is reasonably representative of the group of wells and the property.

This is not the same as adding the wells together, which may cause one or two wells to dominate. Instead, the performance for each well is plotted as starting at the same date or time zero. If the selected wells or groups of wells truly represent similar operational/reservoir characteristics, the collection of individual decline curves should overlay each other or at least demonstrate similar decline form. This approach has been found to work well for the analysis of fractured shale reservoirs such as for the development of models for Marcellus shale production, fault block comparisons in waterflood projects, and for cyclic steam injection projects.

## Type Curves

Fetkovich[8], and others[9], have developed methods of decline curve (rate-time) analysis that are described as **type-curve** analysis. The methods are based on the observation that wells or groups of wells with similar reservoir characteristics will describe similar decline curves over time when compared on a dimensionless time-rate basis. The method requires production history sufficient to demonstrate some depletion but can then be used to determine the form of the decline (including the hyperbolic constant), and certain reservoir properties (assuming other data is available). Type-curve matching makes use of an overlay of actual rate-time data on a set of dimensionless time-rate curves. The dimensionless decline trend which best fits the actual rate-time curve can then be transferred directly to the actual curve to define the future decline trend. The method can employ a number of forms of dimensionless curves. The match points can then be used to calculate certain reservoir properties, such as ($k_h$), using the dimensionless equations. The method has a potential advantage in allowing an evaluator to define a decline form and quantify the hyperbolic constant earlier in the life of a property than would be the case using "eyeball" methods. In theory, it also allows derivation of some reservoir information.

# Use of Crude Oil and Natural Gas Sales Data

Crude oil and/or natural gas sales volumes can be used in place of actual (measured) production reports; however, there are at least five related issues that should be considered before relying on sales data.

First, the methods and accuracy of measurements conducted at the well (or at one or more collection points on a property), as compared to measurement conducted by the purchaser of the oil and/or gas, can differ, resulting in discrepancies. The latter party is measuring a consolidated volume and is paying for that volume; therefore, the purchaser should measure volumes more accurately. This can be more a problem for gas production than for crude oil.

Second, for oil properties, the produced volumes and the sales volumes over a short period of time may not coincide. Oil may be produced and measured but may not be sold until sometime later due to delays in shipping. This is particularly common when production occurs near the end of a month and shipping, whether by pipeline or truck, does not occur until the next month. Production may be reported as occurring in April, but part of the production may not be shipped until early May. Over time (several months or longer), the month-to-month disparities should balance hence the need for data covering a longer period.

Third, the use of sales data must equate the time between sales to the production time. It is not common, but, particularly for small properties, sales may only occur when a certain volume is accumulated, not on a strict calendar basis, so that, over time, the period between sales may increase while the volume appears to be constant. The production rate, however, may be declining.

> As an example, as part of the data set for the evaluation of a group of shallow oil wells in Kentucky, the operator provided a tabulation of production which, when graphed, indicated that there was no decline over a lengthy period of time. The operator later provided copies of the crude oil sales statements covering the same period. It became apparent that the production tabulations were actually sales. The crude oil buyer was only drawing from the storage tanks when the tanks became full. The dates of crude sales indicated that the number of days between sales was increasing. When converted to barrels per month it became apparent that the production was, in fact, declining rather sharply.

Fourth, in certain situations, the produced crude oil may include diluent or other additives. In some regions and fields, crude oil may be difficult to produce due to oil viscosity and composition, requiring the use of light(er) crude or refinery products (such as kerosene) as a solvent or diluent to alter the viscosity of the crude and make it easier to produce. The diluent increases the fluid volume produced and marginally raises the gravity of the oil, resulting in a higher price for the "blended" crude than would be offered for the "native" crude.

In this case, the evaluation must account for the volume of diluent added over a period of time and subtract that volume from the total produced/sold volume to obtain the native crude production. A similar circumstance may occur where NGL is added to the native crude production prior to sale. In both cases it would be reasonable to expect that the operator would maintain records of diluent and/or NGL volumes.

Fifth, there may be another issue related to natural gas production and sales where losses occur due to leakage, or a portion of the gas is used as fuel for the gathering system. The gas sales volume reported from the property may be the produced volume net of the line-losses and/or fuel usage. This raises the question of whether the value of the property should be based on reported production or recorded sales. Losses due to leakage could be considered as a cost of doing business but, assuming there are good wellhead or gathering system volume records, any leakage losses could be measurable and accounted for as an acceptable reduction in value. Use of produced gas as fuel is another story. Fuel gas should be tabulated and projected as an offset to production. The treatment of gas used on the property is discussed under operating costs; however, when the gas production analysis is the objective, all produced gas, regardless of future disposition, should be used.

# Comparison of Production and Sales Volumes

From a practical operating standpoint, there can be an advantage to having access to both the measured production volumes and the sales volumes. As noted above, the volumes may not always be consistent. Comparison of the sales volumes, which logically should be the more accurate, to the reported production provides the opportunity to quantify the magnitude of any differences in reported volumes and to ascertain the source of those differences. Discrepancies resulting from faulty metering and/or leakages in storage facilities or transport lines could be identified and repaired.

# Limiting Conditions

There are a number of issues that must be considered when constructing a production projection for an economic evaluation.

## Physical Limits

| | |
|---|---|
| Reservoir Conditions | – Projections that exceed reservoir capability. |
| | – Unidentified stratigraphic and/or structural interference. |
| Production Capacity | – Projections that exceed the capacity of wells, production equipment, and surface facilities. |

## Rationality of Production Projections

The production projection must be rational. Comparison of the projection to prior production and to similar properties is essential. Particularly when making a projection for an existing producing property, where historical performance data is available, the projection should be compared to the prior performance to either show that the projection is consistent with the demonstrated history OR define where the projection differs from the history. It then becomes necessary to define reasons for the variance and quantify the effects of the difference(s) on the projection of future production.

Comparing the projected production to limiting conditions is essential. When data exists that suggests that there may be a limiting condition such as increasing water production relative to oil production, it may be useful to use historical data to create a relation of water-oil ratio (WOR) to oil production rate and cumulative oil production and investigate and analyze water production as a function of wastewater disposal capacity.

Similar consideration should be given to gas-oil ratio (GOR) as a limiting condition: Would excessive gas production require reducing oil production to avoid wasteful gas production?

## Appropriate Classification of Production

Can the anticipated production be appropriately classified regarding the **risk** of actually recovering the estimated volumes? Particularly where a projection is being made that either exceeds the historical performance of the property or is being made where there is no prior production it is necessary to define the probability of actually recovering the projected volumes. If the projected production exceeds historical performance, there is always some risk that the actual production will be different from the expected volumes. That risk should be accounted for, both quantitatively and by appropriate classification. Projections where there is no prior production carry greater risk of recovery and should be accorded appropriate classification. All projections, regardless of source conditions, carry some risk of recovery.

## Ability to Verify Data Sources

Depending on the source of production data for a particular property, it is often useful to compare the data through more than one source, such as reported production versus sales volumes.

> As an example, it is not unusual for natural gas producers to report actual produced volumes to regulatory authorities, as required by law, but to report only the sales volume to royalty owners.

Another concern, particularly when data is obtained from a commercial or government data source, is the issue of allocation of production among several wells. While this may not be an issue when evaluating a property rather than individual wells, it is important to understand whether the production reported by the source is actual individual well production or is allocated by the operator among the wells. If the latter, it may be prudent to define the allocation procedure.

# Use of Similar or Analogous Production Performance

In the circumstance where the property being evaluated does not have sufficient history to be fully definitive, the **Performance of Similar or Analogous Properties** may be used to estimate producible volumes for a property. If the subject property is a new well, the similar property may be another well or wells on the same lease, or it may be a group of wells on an adjacent or nearby property, or even a group of wells in a nearby field that produces from the same reservoir.

**Similar** or **Analogous** are broad terms and must be used with care. Similar or analogous does not mean identical – no two wells or properties or fields are identical. The properties should, however, share the same reservoir rock and fluid characteristics, as well as operating conditions that would allow an evaluator to expect that the subject property would perform in the same manner as the analogous property over time. Wells in the same field (or portions thereof) and reservoir should perform in a similar manner, whereas wells in different reservoirs may perform very differently. However, even wells in the same reservoir may perform differently if the operating conditions are changed. Analysis and comparison of the performance of a field or lease developed in the 1950s, for example, may not be relevant to a property developed with more modern methods in the late 1990s or the 2010s. The performance of in-fill wells may vary from the original wells if, as is often their purpose, the in-fill wells modify the drainage areas of offset wells.

Single-well comparisons require extreme care, but can be used by imposing the historical performance decline of one well on a new well. Multi-well or family curves combine the performance of several wells by overlaying the decline curve of each well and defining (graphically or mathematically) a composite curve through the set:

1) Select the comparable wells or properties. Define the basis of comparability. This should be more than being in the same reservoir; there should be a comparison of completion method, any known rock and fluid parameters, and method of production.
2) Plot the decline curves for each well or property with initial production points set at a common zero point of the wells that went on production at near the same time or date of first production. Some judgement is needed here.
3) Define a composite decline curve through the graphical set. This can be done by eyeball (it has been done that way before), but computer software to compare variables could be helpful.
4) Use the derived curve to project production for the new well or property.

# Stimulation of Primary Production

The radial flow equation shows that primary production depends upon the pressure differential between the reservoir and the wellbore. A subset of the estimation of primary oil & gas production is the evaluation of properties which are subject to production stimulation techniques. These are methods such as cyclic steaming, fracturing, and acidizing, among others, that alter a portion of

# 6

# PRODUCT PRICES AND PROJECTIONS

The petroleum products that provide value can be grouped as:

(a) Crude Oil

(b) Natural Gas

(c) Associated Products

## Crude Oil

A liquid comprised of hydrocarbon and non-hydrocarbon compounds which, depending on its composition, has differing properties including oil gravity, viscosity, sulfur content, and pour-point. These latter properties, at least in part, define the quality of the oil for potential end-use and also strongly influence the methods that could be used to develop and produce the oil. Crude oil that has a high API gravity and low fluid viscosity, such as West Texas Intermediate (WTI), is generally easier to produce than low-gravity, high-viscosity oil, such as Midway-Sunset 13° API; the latter may require some form of stimulation in order to maintain efficient production rates. The characteristics of the oil can, and often do, influence everything from well spacing to pump size to life of production – each of which has an impact on the economic benefit of the producing property.

Crude oil is a market commodity and is subject to considerable variation in its economic value based on location, gravity, sulfur content, and competition from other fuel sources.

As an example, crude oil produced in California is consistently priced several dollars per barrel below oil produced in West Texas or in the Mid-Continent due in part to the large difference in oil gravity but also because of comparatively higher sulfur content as well as restricted access to other markets; virtually all California crude is refined locally.

## Natural Gas

A fluid also composed of many hydrocarbons and, occasionally, non-hydrocarbon compounds. The primary differences between gases are the heating value (generally the methane content), the non-hydrocarbon gas ($N_2$, $CO_2$, $H_2S$, etc.)

content, and the amount of liquid that can be obtained as natural gas liquids (NGL). The economic value of natural gas is directly related to its composition. A gas comprised primarily of methane with no non-hydrocarbon content has a high heating value. If it contains natural gas liquids, the economic value often increases. Natural gas enjoys a relatively high economic value because it has a low(er) cost of production and processing compared to crude oil. However, it is highly regulated through pipeline and utility controls on transportation and pricing. In addition, due to the relative ease of pipeline transportation of gas, where pipeline access exists, there is substantial competition among gas-producing regions which tends to act as a control on gas prices.

## Associated Products

Include natural gas liquids (NGL), sulfur, and non-hydrocarbon gases. NGL is most commonly found in natural gas that is produced in association with crude oil and can have substantial economic value but pricing tends to be driven by local markets and can be volatile with demand. NGL is generally treated as a byproduct of gas sales where wet gas containing NGL is sold to a "gas plant" which extracts the liquids and then sells, or otherwise distributes, the "dry" gas and the derived liquids separately. The producer may receive either revenue from the dry gas sales and a share of NGL sales or a "wet" gas price. If the petroleum producer is also operating a "gas plant" and is removing NGL from the gas, the liquids may be mixed back into the produced crude oil to raise the gravity of the oil prior to sale. On other properties, NGL may be segregated into ethane, butane, propane, et al., which can be projected and priced separately if a definable and consistent market(s) exist. The economic treatment of NGL and other by-products depends upon the facilities available to the property and the existing market.

Sulfur is a by-product in many Canadian fields and in some areas of the United States. The economic value of the sulfur content tends to fluctuate considerably, and sulfur production may often be more of a nuisance than an economic benefit. Non-hydrocarbon gases such as nitrogen, carbon dioxide, and helium can be economic by-products; in some cases where these gases occur, the non-hydrocarbon gas may become the primary product.

## Product Prices

Discussion of petroleum economic parameters usually centers on the price paid or offered for the crude oil and natural gas produced. Some property evaluators consider current price as well as expectations of future prices to be the most important parameter in an evaluation. This is a debatable premise. Product price is certainly one of the most visible parameters if for no other reason than that it is reported in the media day and night and is a continual source of theorizing and speculation. Product price is potentially the most volatile component, with changes occurring monthly, weekly, daily, and in some cases, hourly. Product price therefore deserves appropriate analysis before application to an evaluation. Consideration of oil & gas prices is important in two areas of petroleum economics. First, product price is a direct and important component of individual property evaluation, as we will discuss further. Second, studies have shown that oil (and, to a lesser extent, natural gas) price and the changes thereof, are an important, if not the most important, short-term driver of petroleum company economic returns and stock prices. Leaving corporate concerns to financial and academic experts, we will concentrate on the property evaluation aspect.

The choice of the initial price and the projection of price into the future can have a critical impact on the evaluation of a property.

Important caveat: There is no correct initial price or price projection, only, theoretically, a rational one.

A brief perusal of industry and government literature will show that estimates of future prices change almost daily. This condition does not mean that an evaluator is free to adopt any price scenario that suits him. Projection of a rational income stream requires an *estimate of future product prices that is consistent with the economic conditions that are likely to prevail over the duration of the projection.* The following section will discuss theories and practical methods for deriving initial or starting prices for oil and/or natural gas and defining projections of those prices into the future.

# Product Prices as a Function of Available Markets

## Crude Oil

Crude oil is a global commodity subject to constant influences of supply and demand, augmented, and in some cases overridden and/or driven, by government interference and regulation. A modest interruption of supply (or demand) in one part of the world can, and often does, cause oil prices in markets everywhere to change. Further, the impact can often be highly variable, suggesting and confirming that oil prices can be volatile over the short term.

The history of crude oil prices, both domestically and globally, has been discussed and debated in an uncountable number of books, magazine and newspaper articles, government and private studies, learned academic treatises, as well as informed (and not so much) opinion commentary, so it is not necessary to reprise the subject here in any detail. Two graphical studies should be sufficient to provide background.

One of the more interesting and informative depictions is Figure 31 that shows crude oil prices in both nominal and real terms since January 1946. In this study, the price is for Illinois sweet crude using data from Plains All American Pipeline through November 2022, and nominal prices (black) are converted to real prices (red) using November 2022 as the base.[12] There is a lot of information presented here that illustrates the changes in price over a 76-year period incorporating the pre-OPEC proration era.

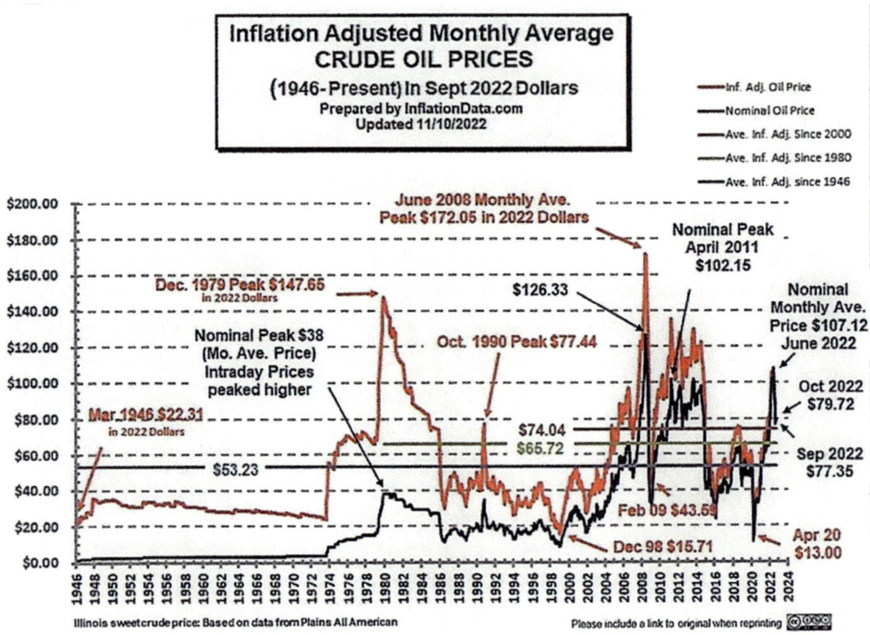


**Figure 31.** Inflation adjusted monthly average crude oil prices. (inflationdata.com)

As shown in the graph, there have been several periods when oil prices increased to temporary peaks: in December 1979, June 2008, April 2011, and most recently, in June 2022. Also of interest is the analysis presented in the graph of average inflation-adjusted prices over certain periods. Since 1946 through November 2022, the average *real* price has been $53.23 per barrel. In contrast, the *real* price average since 2000 has been $74.04, a roughly $20 per barrel increase over a period of 54 years.

Figure 31 does not incorporate any information regarding economic or political/regulatory events that could affect price. Another view is shown in Figure 32 which presents data from the U.S. Energy Information Agency (EIA) based on West Texas

Intermediate (WTI) crude oil prices from 1986 through 2022 and includes selected information on economic, political, and regulatory issues. It should be apparent that, to a large extent, oil prices are less affected by supply/demand issues than by political/regulatory events, BUT those events can (and usually do) create supply/demand interruptions that in turn cause changes in crude oil prices. On the other hand, some supply/demand events can result in political/regulatory reactions. The expansion of oil production from hydraulically fractured shale reservoirs in the early 2000s increased the volume of oil (and natural gas) available in the United States and temporarily made the United States the world's largest oil producer, regaining the title lost in the early 1970s. The downside of the extensive increase in oil production was a decline in oil prices in 2014-15.

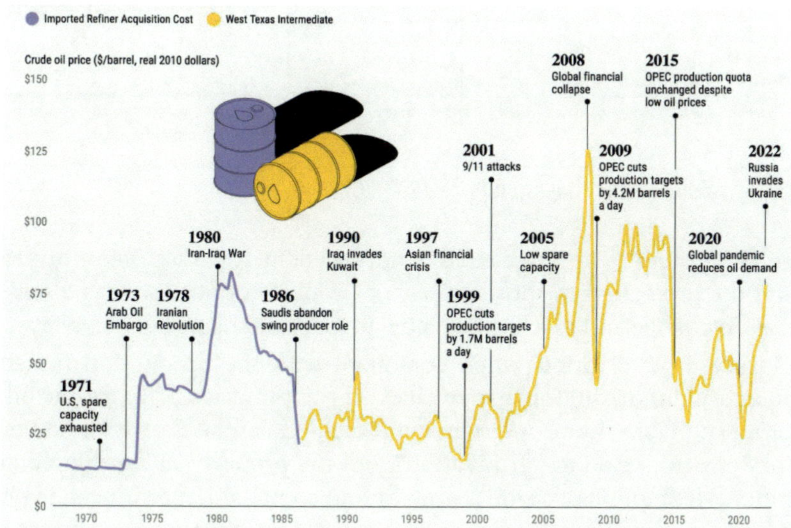


**Figure 32.** Historical oil price reaction to global events. (U.S. Energy Information Administration, Refinitiv (March 2022))

A similar graph can be shown for natural gas prices (Figure 33), which shows a wide range of prices for natural gas from January 1997 through late 2022. As noted, the market for natural gas tends to be more local or at least regional than for oil and to reflect local conditions such as weather and regulatory/political controls.

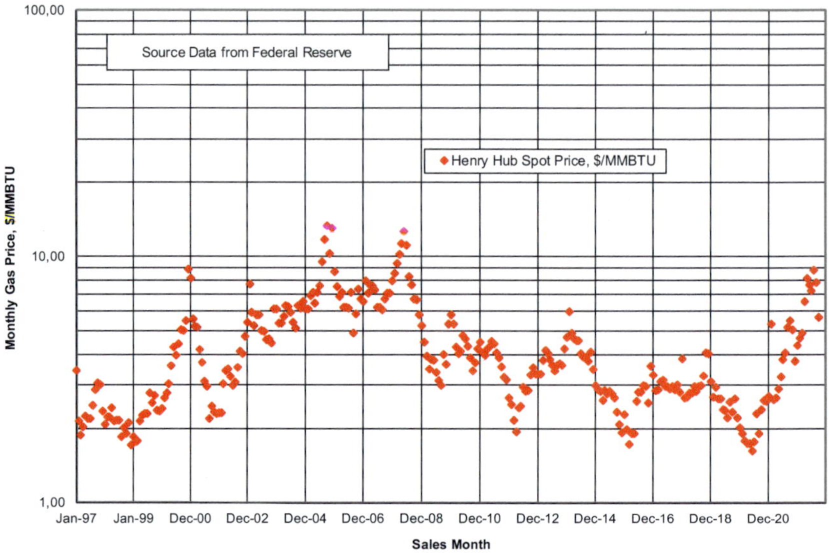


**Figure 33.** Henry Hub Natural Gas Spot Price 1997-2022

These few examples are presented as a precedent to a discussion of product prices and the projection of those prices for property evaluations, as a reminder that prices do change and that the potential for such changes must be considered.

The future product price, when combined with the production projection, results in a revenue stream for that product. The appropriate price for the oil and/or gas produced from the property being evaluated can be derived by reviewing the history of prices paid for production from the property, and/or by obtaining one or more prices for oil of a similar gravity and composition produced in or near the same field, and making necessary adjustments for gravity and other relevant factors.

# Initial Crude Oil Price

An income stream is usually constructed as of a specified effective or "as-of" date; therefore, the initial oil and/or gas price should be representative of the price in effect at that date. Depending on the purpose of the evaluation, the selection of an initial oil price may be relatively uncomplicated. In the case of internal evaluations for budgeting or capital expenditure analysis, management may dictate the oil and/or gas price(s) to be used.

For some types of evaluations, the initial price may be specified as the price paid or offered on a certain date, such as January 1 (for property tax), or, depending on jurisdiction, for inheritance tax. Or, it may be the average of prices actually paid or offered over a certain period preceding the effective date. In some circumstances the definition of the initial price may be left to the judgement of the evaluator. The selection of an initial oil price is not necessarily an easy decision, but requires a knowledge of available information sources followed by considerable data collection, interpretation, and analysis.

## Product Sales Statements

The most reliable source(s) of information, when available, is the **price actually paid** for production from the property over a period of time and/or the prices paid for similar production in nearby fields prior to, but in close proximity to, the effective date. The existence of large-scale property- and field-wide computer-monitored shipping and transportation reporting systems should make the accumulation of sales and related data efficient and readily available in digital format. Such systems are often not available to small operators, which then requires collection, review, and analysis of actual (hard-copy) sales records. Sales statements commonly report transactions occurring within a calendar month. The format and content of the statements may vary but are particularly useful if they come from a third-party crude purchaser rather than an internal transaction between divisions of a company. Three or four years of monthly sales history will commonly provide sufficient data to provide insight into the revenue foundation of the property. A detailed sales statement will commonly provide data such as:

- Volume of oil sold
- The gravity of oil sold
- The total amount paid for the oil
- The oil price (calculated as amount paid divided by volume sold)
- Sales point (LACT unit, pipeline connection)
- Owner interest share being paid (working interest, royalty interest)

The sales statement may also provide data regarding charges imposed by the buyer or transporter which can be calculated (individually or collectively) as a unit cost/charge in $/barrel:

- Shipping of oil (truck, pipeline, rail), $
- Dehydration, $
- Heating of oil for transportation, $
- Diluent Usage, $ (see below – Diluent Usage)
- Severance and other state and/or local taxes, $
- Any other allowed deductions from sales revenue

As an example, oil sold from a property and shipped by truck may show a price paid for the crude but may then also show a deduction for the trucking charge. These charges must be accounted for in the income stream. The detail reported on the sales statements should allow calculation of the various deductions so that they may be applied to the evaluation. Care must be taken to ensure that the oil price derived from the sales statement is the gross price before any deductions. This can usually be determined by comparing the net amount paid to the sales total.

If there is a drawback or limitation to the use of actual sales, it is that the reported price is a single point indicator which may be only an average for the sales period. When data access is possible, the monthly sales price should be compared to sales or transactions that occur during the month covered by the sales statement. For instance, if there were four transfers during the month that received different prices, comparison of the **volume weighted price** to the reported month average may be useful in extracting a representative price.

## Posted Prices

A second source for initial price is the **posted price** for oil from the property. Posted prices are the prices offered in the market by large purchasers, such as major refiners, for crude oil from a particular field and are usually contained in a published "posting" list. The posting may be effective for a period of time but may also change from time to time as economic conditions vary. Posted prices are readily available online for oil from various fields or for regions of the country or state.

The example below (Figure 34) is a Chevron North American Oil Marketing Summary posted price Bulletin for California for December 2022. Two immediate observations can be made. First, prices are posted for two fields, Buena Vista Hills and Midway-Sunset, both located in Kern County, California. The posted price is listed as $/Bbl and is a function of the oil gravity. The base gravity for Buena Vista Hills field is 26° API, while Midway-Sunset base gravity is 13° API. There is a gravity adjustment, as shown at the bottom of the Bulletin, from the base price for increments Above 40° API, 34.0° to 40.0° API, and Below 34° API. The gravity adjustments are measured in $/Bbl/Tenth ° API.

Second, there are separate prices posted for Buena Vista Hills and for Midway-Sunset for twenty-one (21) days in December and a weighted average for the month. In theory, and generally in practice, oil sold from Midway-Sunset on December 7 would receive a price of $73.51/Bbl. Oil that is collected by truck or shipped through a LACT unit on December 20 would be priced at $76.59/Bbl. Oil transferred on December 30 would receive $80.51/Bbl. The weighted average posted price for all of December 2022 is $82.2968 for Buena Vista Hills and $77.7468 for Midway-Sunset. The price difference is almost entirely related to the gravity differential.

Assume that the property being evaluated produces 21° API crude and that the oil is similar in all other characteristics to Buena Vista Hills. The gravity differential between the subject property and Buena Vista Hills is 5° API (50 basis points). The gravity adjustment is 50 times $0.0350/Bbl which equates to $1.75/Bbl. The average Buena Vista Hills price for December of $82.2968 would be reduced by $1.75 to $80.5468 for 21° API crude from the subject property. This may be sufficient information for application to an evaluation, but it should be noted that postings are made, often daily, whenever there is a change in the oil market.

The Chevron posting for Buena Vista Hills changed twenty-one (21) times in December from a high price of $87.77 on December 1 to a low of $76.57 on December 9, 2022; the posted price then increased over the remainder of the month to $85.06 on December 31. The Chevron average for December is apparently daily weighted. This may suggest that the monthly average may not be representative of the initial price. If there is a strict requirement to use the price on December 31, for example, the average price may not be acceptable. It may also be useful to note that the Chevron average posted price for calendar year 2022 ranged from $85.6568 in January to $82.2968 in December, with a peak price of $119.7707 in June 2022.

The Chevron posted prices provide information about the price of oil for Buena Vista Hills and for Midway-Sunset, but it may be useful to obtain posted price information from more than one source for comparison to information from sales statements. For example, three other companies – Plains, Shell, and ExxonMobil – post prices for Buena Vista Hills 26 and for Midway-Sunset 13. Figure 34 below indicates some differences between the four companies in each month but a considerable consistency as well. This should be expected for companies competing for the purchase of crude oil.

Chevron

Crude Oil Marketing

**North America Crude Oil Marketing - January 2022 Summary California Bulletins 22-CA1 to 22-CA9, with Monthly Averages**

Chevron has changed its posted prices as shown below:

Effective 7:00 AM on the above listed effective date, subject to change without notice, Chevron will pay the following prices per barrel for crude oil and condensate delivered for its account into the custody of its authorized carrier or receiving agent. Prices are subject to deduction for transportation and other charges where applicable. Prices for oil apply to quantities measured by approved automatic custody transfer facilities or 100% tank table with customary adjustment of volume and gravity for temperature and full deduction for sediment and water.

* Prices are subject to a Gravity Adjustment Scale

| California | API | CA1 01/03 | CA2 01/04 | CA3 01/05 | CA4 01/06 | CA5 01/07 | CA6 01/10 | CA7 01/11 | CA8 01/12 | CA9 01/13 | CA10 01/14 | CA11 01/18 |
|---|---|---|---|---|---|---|---|---|---|---|---|---|
| Buena Vista Hills | | 79.63 | 80.04 | 80.90 | 82.51 | 81.95 | 80.41 | 84.27 | 85.69 | 85.17 | 86.87 | 88.48 |
| Midway Sunset | | 75.08 | 75.49 | 76.35 | 77.96 | 77.40 | 75.86 | 79.72 | 81.14 | 80.62 | 82.32 | 83.93 |
| Gravity Adjustment | | Price Adjustment $/bbl/Tenth°API | | | | | | | | | | |
| Above 40.0°API | | 0.0000 | 0.0000 | 0.0000 | 0.0000 | 0.0000 | 0.0000 | 0.0000 | 0.0000 | 0.0000 | 0.0000 | 0.0000 |
| 34.0° to 40.0°API | | 0.0100 | 0.0100 | 0.0100 | 0.0100 | 0.0100 | 0.0100 | 0.0100 | 0.0100 | 0.0100 | 0.0100 | 0.0100 |
| Below 34.0°API | | 0.0350 | 0.0350 | 0.0350 | 0.0350 | 0.0350 | 0.0350 | 0.0350 | 0.0350 | 0.0350 | 0.0350 | 0.0350 |

| California | API | CA12 01/19 | CA13 01/20 | CA14 01/21 | CA15 01/24 | CA16 01/25 | CA17 01/26 | CA18 01/27 | CA19 01/28 | CA20 01/31 | | Mthly Avg |
|---|---|---|---|---|---|---|---|---|---|---|---|---|
| Buena Vista Hills | | 89.51 | 89.20 | 87.44 | 86.36 | 88.65 | 90.40 | 89.66 | 90.12 | 90.95 | | 85.6568 |
| Midway Sunset | | 84.96 | 84.65 | 82.89 | 81.81 | 84.10 | 85.85 | 85.11 | 85.57 | 86.40 | | 81.1068 |
| Gravity Adjustment | | Price Adjustment $/bbl/Tenth°API | | | | | | | | | | |
| Above 40.0°API | | 0.0000 | 0.0000 | 0.0000 | 0.0000 | 0.0000 | 0.0000 | 0.0000 | 0.0000 | 0.0000 | | 0.0000 |
| 34.0° to 40.0°API | | 0.0100 | 0.0100 | 0.0100 | 0.0100 | 0.0100 | 0.0100 | 0.0100 | 0.0100 | 0.0100 | | 0.0100 |
| Below 34.0°API | | 0.0350 | 0.0350 | 0.0350 | 0.0350 | 0.0350 | 0.0350 | 0.0350 | 0.0350 | 0.0350 | | 0.0350 |

Note: Prices in $/bbl

**Measurement**

The prices are based upon computation of quantities by mutually acceptable automatic measuring equipment in accordance with the latest ASTM Standards and API Manual of Petroleum Measurement Standards (MPMS) generally accepted in the petroleum industry with deductions for full sediment, water and other impurities.

**Figure 34.** Chevron North American Oil Marketing Summary posted price Bulletin for California. (chevron.com)

**Weighted Average Posted Prices for Midway-Sunset 13° API – July through December 2022**

| | Chevron | Plains | Shell | ExxonMobil | Average |
|---|---|---|---|---|---|
| Month | $/Bbl | $/Bbl | $/Bbl | $/Bbl | $/Bbl |
| July | 103.8429 | 99.6655 | 102.0952 | 104.5474 | 102.5376 |
| August | 94.9261 | 92.5539 | 94.6323 | 96.1384 | 94.5627 |
| September | 88.2743 | 84.5030 | 87.0817 | 88.9410 | 87.2000 |
| October | 91.3397 | 86.8394 | 89.5129 | 92.1077 | 89.9499 |
| November | 88.7050 | 84.4950 | 85.5217 | 89.5483 | 87.0675 |
| December | 77.7468 | 73.4771 | 74.8500 | 78.6165 | 76.1726 |

Use of posted prices has a limitation in that there is no information regarding allowable deductions. The price that is realized on a specific property may be subject to charges such as trucking, compression, pumping, diluent separation, and dehydration that do not apply to other properties in the posted field.

## Comparison of Sources

While obtaining and analyzing both sales statements and relevant posted prices will require additional work, there can be value in the effort. Having two sources allows comparison of the reported sales price to the posted price. As noted above, the price from the sales statement is usually a single-point average, dependent on the timing of the sale and the prevailing price. On the other hand, the posted price may vary considerably over a month so that the daily weighted average may be different from the reported sales price. The variation of the posted price over a month or several months may indicate that the historical and potential price for oil is more volatile than indicated by the reported sales price.

The actual price paid for crude oil (before any deductions) may be equal to, or very close to, the posted price if the property is located in a "posted" field or area and has a gravity similar to the posted gravity. Otherwise, the posted price can be used to estimate the price of oil for a particular property after adjustments are made for location and gravity. There may also be a difference between the actual price and the posted price if the oil varies in some characteristic from the benchmark crude, such as higher sulfur content or other compositional difference; additional corrections may be necessary. Lastly, a comparison is useful, if possible, especially when prices are changing frequently because sales may only be reported as a total for previous month or two, whereas posting often occur several or more times in a month.

The actual price may differ from posted price because:

- the oil is not being sold to a buyer who posts prices for the field or area,
- the crude oil sales contract may specify that the purchaser may deduct certain costs when and if necessary or pay a premium or deduct a certain amount due to crude quality, or
- the terms of the operating agreement and/or mineral rights lease require payment based on other than actual sales or posted price. Some operating agreements or oil & gas leases require that payments be based on an average of posted prices from specific sources rather than the actual price paid.

## Crude Oil Purchase Contracts

A third source of price information is an existing contract for the purchase of crude oil from a property by a pipeline and/or refiner. Such contracts are not uncommon, particularly for properties that may be remote from other fields or from transportation facilities such as pipelines, railroad access points, or trucking services and where the purchaser must invest in such facilities. In this case, the producer (seller) may be guaranteed that production will reach a market, while the purchaser is assured of a crude source. The purchase contract will usually include a price mechanism such as equating the price to posted price(s) for a similar field or an average of the posted prices for a similar field from multiple posting sources. The contract may also include a description of the charges that may be deducted from the sales revenue by the buyer. Assuming that the buyer has made some investment in transportation facilities, the contract may have a long duration. Where such a contract exists, the terms should be considered the primary source of price data augmented by the actual sales history.

## Representative Price

Evaluation as of a given date is useful for standardizing and comparing results among a range of property valuations for capital budgeting or other internal purpose. Evaluation as of a given date may also be required for certain regulatory purposes (i.e., estate tax, ad valorem tax). However, in neither case does the evaluation necessarily require the use of the crude price that exists on the effective date. The goal of evaluation is to derive a value that is representative of the value of the property on (or about) that date. The Chevron Posted Price example of twenty-one price changes in one month (December 2022) suggests that some flexibility is needed.

Assume a property tax evaluation done as of December 31. One approach is to use the oil price actually paid on the date of evaluation as the initial oil price. This approach is simple, direct, and removes the need for judgement and debate. However, as previously noted, oil prices change frequently and, on occasion, by large amounts. The price on a specific day may or may not be representative of the value of the crude oil on or about the evaluation date. Using the Chevron posted price example, the weighted average price for Midway-Sunset crude in December 2022 is $77.75/Bbl, but the price on December 30, the last posted date, was $80.51/Bbl, which is $1.83/Bbl higher than the price on December 29 ($78.65/Bbl) and $8.05/Bbl higher than on December 8, 2022 ($72.46/Bbl). The differences are important and, depending on which price (or prices) were used as the initial price, could result in an exaggerated or understated property value. In this case, use of an average for the month may be more representative of the actual price.

Other options, which tend to be less formalized, attempt to account for changes in oil prices over time by using an average of the prices over the previous 3, 6, 12 months, or some other period prior to the evaluation date. It could be that the 3-, 6-, or 12-month averages are relatively close to each other and, better yet, to the December average. These approaches focus on deriving a representative price to be used as a starting price. This may not be the same as the effective date price. Using the Chevron example, if the effective date is December 31, the posted price for Midway-Sunset 13° is $80.51/Bbl on December 30, but the average for all of December is $77.75/Bbl or $2.76 per Bbl lower than the year-end price. The average price over the last three months (December, November, October) is $85.93, which is higher than both the December 30 price ($80.51/Bbl) and the December average price ($77.75/Bbl); the six-month average is even higher at $90.81/Bbl. The downtrend in price suggests that a more in-depth review is needed to ascertain if the six-month average is an aberration or part of a long-term shift in prices.

An evaluator should always be free to exercise judgement and to use another price or a range of prices subject to appropriate documented support for the choice; all that is necessary is that a rational support for the price selection be provided. For required valuations, such as for ad valorem or estate tax, change should not be arbitrary. If the evaluation is required to be based on a mandated effective date, such as December 31, then the price on that date should be considered as an indicator of the starting price. However, it may not necessarily be used if an alternative price can be supported with a rational explanation.

The Energy Information Administration (EIA), among other public and commercial sources, provides a large amount of continually updated and readily accessible data regarding natural gas production, transportation, usage, markets, and prices that can be useful in the evaluation of natural gas properties.

## Natural Gas

Natural gas is an energy commodity that differs from crude oil in several ways. Natural gas prices were gradually deregulated through the 1980s; full wellhead price deregulation came in the 1990s. Pricing of natural gas is determined by supply, demand, and transportation (primarily pipeline) factors. Unlike crude oil, natural gas is used directly in virtually all sectors of the economy including residences, businesses, industry, and power generation. Natural gas is more homogeneous in composition but is more difficult to transport. As the result, trade in natural gas is closely tied to the availability of pipeline capacity.

Natural gas is produced in two forms: **"dry" gas** and **associated gas**. The difference is that associated gas is produced along with crude oil from oil reservoirs, whereas dry gas is produced from reservoirs that contain only gas (such as the Hugoton field or the modern Marcellus, Haynesville, or Fayetteville "fields"). Both forms have been a source of some revenue as long as oil has been produced, but the economics of the two forms have different histories. Dry gas could be produced and sold in local markets or in some instances pipelined to distant markets. On the other hand, associated gas was not accorded much economic value until the increase in natural gas applications for residential and commercial use in the latter half of the 20th century coupled with environmental regulations improved the marketability of natural gas. Gas that was produced along with crude oil was commonly either used as fuel on the property (UOL) for operations or was burned ("flared"). As natural gas assumed greater market value, flaring was reduced and other sources of energy such as electric power were substituted for produced gas. There was also a concern, particularly regarding solution-gas-drive oil reservoirs that, as oil production declined, the relative volume of produced gas would increase and, absent a means of removing the gas, oil production might be restricted. The result was that associated gas had to be accommodated. Fortunately, the increasing market for natural gas created an outlet and added a revenue stream to many properties.

## Initial Natural Gas Prices

The above discussion has been directed toward selecting an initial, representative oil price. The principles of deriving a representative price apply just as well to natural gas pricing but with a few additional considerations. Natural gas is not as transportable, world-wide or across large regions, as is crude oil or other liquid products. Absent investment in long distance pipelines and/or liquefaction facilities, the market for gas from a particular area or region may be limited. Gas sales are not always as straightforward as oil sales. During periods of high demand, the purchaser may be willing to take all the gas that can be produced and will build the collector lines to the property. At other times, however, the purchaser may limit sales to a percent of capacity and may not be willing to provide connection. These conditions could not only reduce revenue from sales but could require additional investment to build a collection/transportation system and, further, in the case of associated gas, may have the effect of reducing production of crude oil and other products. Present and future anticipated market conditions must be considered in constructing both the production schedule and the revenue stream.

Further, gas is often subject to non-market or extra-market influences, such as government and/or regulated utility rule-making, which must be considered in an evaluation. Natural gas, whether dry or associated, is commonly sold from a property into a **collection** or **gathering system** which takes the gas from the property (and probably other properties) to a local pipeline for connection to a regional pipeline. The connection point of the gathering system and pipeline and/or the connection of one or more pipelines is called a **hub.** The local price at any other location or point of consumption in the commodity network is the hub price plus the cost of transportation from the hub to the point of consumption. Price differences between any two points in the pipeline network represent just the cost of transportation between those points. Absent other sources, these "hubs" may often be the control point for the gas price.

There can be an important difference in price based on the means of transporting gas to a regional pipeline or other consumption point.

> As an example, a property in north central Pennsylvania produces Marcellus shale gas from two groups of wells. Group A, with six wells, was drilled and produced by the current operator (XYZ Company). Group B, with four wells, was drilled by another company. The Group B wells were acquired by XYZ. In addition to drilling the Group A wells, XYZ also built a gathering system that transported

the produced gas a considerable distance to a regional pipeline. The previous owner of the Group B wells did not build a gathering system and concluded a sales contract that relied upon the pipeline to collect the gas at the wellhead and transport it to a sales point. The Group A wells receive a gas price that is $0.65 higher than the Group B wells. The difference is the cost of transportation.

## Product Sales Statements

As with crude oil, review and analysis of monthly gas sales statements collected over a period of 2-3 years is usually the best source of information regarding the price of gas for a specific property. Use of product sales statements for natural gas follow the same precepts as using sales statements for crude oil. Gas sales prices are most commonly based on the heating value of the gas measured in millions of British Thermal Units (BTU), not on the volume of gas. Since gas production is normally expressed in MCF, a correction must be made if the gas price is quoted or reported in dollars per million BTU ($/MMBTU). Natural gas sales statements often provide data regarding:

- Volumes of gas sales (MCF and/or MMBTU)
- Total Sales, $
- Sales price ($/MCF and/or $/MMBTU)
- BTU content (MMBTU per MCF)
- Sales points
- Owner interest share being paid (WI, RI)
- Information regarding any deductions from sales revenue.

The ability to use natural gas sales data in comparison to gas production for a property over a period of time can be useful for the same reasons as was noted regarding comparison of crude oil sales and production. Anomalies can occur due to differences in measurement at the wellhead and in the gathering system and at the pipeline connection, differences between the production period and the sales period, and the use of some produced gas on the property. In the latter case, a portion of the gas is not sold but may be retained for use as fuel, reported as **Gas Used-on-Lease (UOL)**, which can vary with the season. Since this UOL volume is not sold, it will not be reported on the sales statements; the UOL volume must be determined separately and, if necessary, accounted for in the income stream.

## Gas Sales Contracts

Natural gas sales differ from crude oil sales, to some extent, in that gas is often purchased under some form of contract to a pipeline or end-user covering a specified period of time, which bears a specific price and time period, and may be subject to various deductions. Pipeline contracts may be of short duration (1-year), but end-user contracts can be for multiple years and may also be for specified volumes of gas. End-users may be a processing facility such as a fertilizer or chemical plant, an electric utility, or a commercial complex that wants to be sure of a supply of gas over a longer period of time. In such cases, the contract pricing is defining for the period of the contract.

Many properties continue to sell gas under contract to third parties such as pipelines and/or utilities; the contracts tend to be shorter and may often have pricing formulas that require action by third parties unrelated to the producer or buyer. In some areas, gas is a true commodity subject to supply and demand and offered for sale to the highest bidder. In other cases, gas is sold into the spot market or sold in short-term volume deals to specific end-users. The "spot-market," operating through dealers or clearing houses, offers some of the aspects of "posted" price for gas, but, as the term implies, is more transitory than the oil price postings and is not readily obtained. Spot prices can change numerous times a day.

In evaluating a property with a long-term gas sales contract, it is essential to understand how the terms of the contract affect the price paid for gas. If the price is stated and fixed for a specific period, the contract price should be used. When evaluating a property where gas is sold into a spot or other transitory market, some determination of an appropriate initial price must be made based on the actual experience of the property in selling gas and on the expectations of the marketplace.

An argument is occasionally advanced that the contract price should be ignored in favor of a "market" price that could be otherwise obtained. If the contract is specific to the property and would be transferred to a successor property owner, the contract price is the appropriate price to use. If not, then it could be assumed that the new owner (seller) would be able to obtain market price. It should not be assumed that the new property owner is a superior negotiator able to obtain the best possible contract, but rather is able to obtain a price and terms similar to those in effect for other properties. As in all other issues, the purpose of the evaluation may mandate the choice of the initial price, either by regulation or custom.

## Sales Hub Prices

There is no readily accessible system of *posted prices* for natural gas as there is for crude oil. However, natural gas prices can be estimated from prices reported for local and/or regional connection "hubs." As noted above, natural gas is produced on a property and collected into a gathering system, which is then connected to (possibly) a larger gathering system before being transferred to a pipeline for transmission. As previously mentioned, the point of connection of the gathering system to the pipeline is referred to as a hub. Figure 35 shows the location of some, but not all, of the hubs in the northeast United States. There are at least two online commercial sources that identify these hubs; both sources list prices being paid or offered at these hubs on a daily basis along with other information regarding natural gas markets.

User caveat: These are commercial subscription sources that provide a large amount of data, but they are expensive.

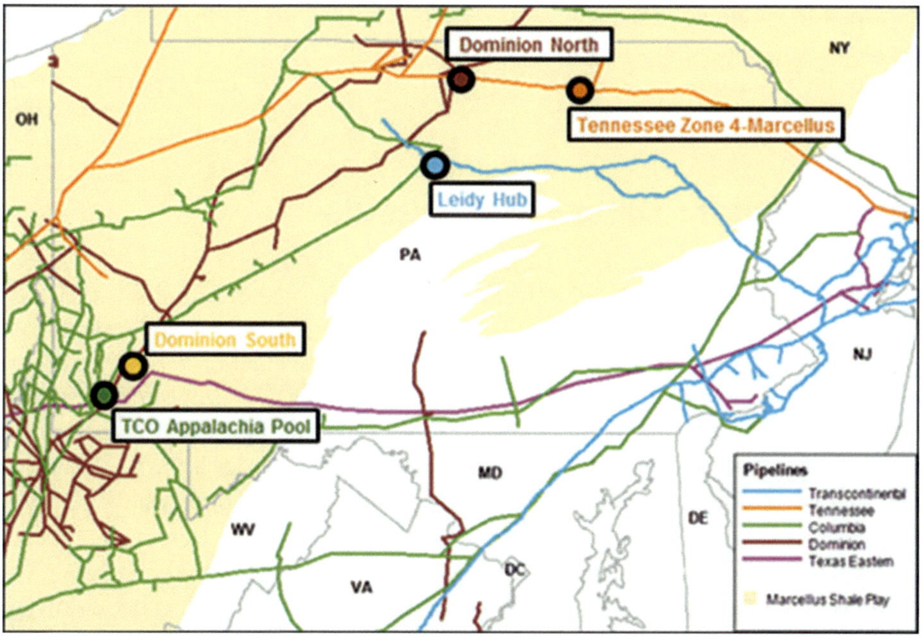


**Figure 35.** Northeast United States Hubs (U.S. Energy Information Administration)

One online source lists a total of 172 hubs, of which 24 hubs are in the Northeast and another 13 hubs are in the East Texas region. These data sources provide similar information in slightly different formats. Both sources provide a daily report of average gas prices at each hub along with the volume of gas at the hub (MMBTU/day) and the number of gas sales. This level of price information could be used as a source of historical prices for a property **IF** the appropriate destination hub was known. The information could be used in a manner similar to crude oil posted prices as a means of comparison to actual sales prices and/or where sufficient sales data are not available.

## Representative Price for Natural Gas

The process of deriving a representative gas price for an evaluation is no different than deriving a representative oil price. The initial gas price can be the actual price on the effective date or the average of prices over a period prior to the effective date. The only requirement is to obtain a rational and documented representative price to be used as a starting point. An evaluator should always be free to exercise judgement and to use another price or a range of prices, subject to appropriate documented support for the choice; all that is required is a rational support for the price selection to be provided. For required valuations, such as for ad valorem or estate tax, change should not be arbitrary. If the evaluation is required to be based on a mandated effective date, such as December 31, then the price on that date should be considered as an indicator of the starting price. However, it may not necessarily be used if an alternative representative price can be supported with a rational explanation supported by appropriate data.

## Initial Price for Natural Gas Liquids

Deriving an initial price for Associated Products should follow the same procedures as for crude oil and natural gas, where the use of actual sales data is the primary (and probably only) source of extraction and pricing data. Products for which a market exists should be included in a property evaluation but only if there is either a demonstrated history of sales from the property and/or similar nearby properties, and/or a confirmed market for the product along with a commitment for development of necessary facilities. This work is made easier if the NGL is sold to a third-party processor which provides processing and sales data records.

The best source of data is the actual production and sales of each product over a reasonable period prior to the effective date; the longer the period, the better. In some cases, such as for NGL, it may be necessary to determine product breakdowns to propane, butane, etc., to see if price variation occurs due to seasonal demand. This may require a substantial amount of work that may or may not be reflected in the quality of the evaluation. The efficient approach may be to aggregate gas sales and NGL sales into a wet gas price to be applied to sales gas volume. On the other hand, if NGL is separated from the gas and then mixed with the crude oil to improve the price obtained for the oil, then the realized oil price will reflect the NGL content. In this instance, it may be necessary to determine whether the production of NGL varies with time and the extent to which the crude oil sales price is modified by the addition of the NGL. The actual experience on the property should be considered definitive.

If there is a demonstrated market for NGL or other products, such as by sales from nearby similar properties, but no history of such sales from the subject property, an evaluator could consider that the future sale of those products is a reasonable expectation and could project such sales. Extraction of NGL or other products from produced wet gas requires separation facilities either on the property or, more commonly, at an off-site location where the facility may serve numerous properties and operators. In addition, there must be shipping facilities to distribute the products either back to the source properties, to sales points, or both. Such facilities can be large investments. There must also be a provision in the evaluation for including the capital expenditures and additional operating costs necessary to establish such sales.

Depending on the scale and function of the facility relative to the subject property and, possibly, other properties, it may be useful to treat the facility as a separate operation with a production and income stream distinct from but connected to the economics of the property.

# Estimating Future Product Prices

An Income Approach evaluation requires an estimate of future revenue from the property which, as a precedent, requires an estimate of **future** product prices. Having said that, the evaluation should be done with the realization that hydrocarbon prices are subject to frequent changes and any projection of prices for oil, gas, or other product in the absence of incredible prescience will be wrong.

The projection of product prices for 5, 10, or 20 years depends on the perspective of the evaluator regarding future economic conditions and, to some extent, the purpose of the evaluation. Major companies often have economics departments whose purpose is to estimate future prices, which are then incorporated into income streams. Smaller companies and individual interest owners are generally without such resources and must rely on published and unpublished sources, third-party studies, and/or their own experience and intuition.

The simplest approach is to determine the representative price(s) as of the effective date and hold that price constant for the life of production. This was virtually the only way projections were done prior to the price increases of the 1970s, and the approach is still in use. Since that time, it has become more common to attempt to estimate the extent to which oil and/or gas prices could change over the expected life of production and to build those anticipated changes into the income stream. In general terms, this has been done by escalating or de-escalating oil prices at certain fixed or variable rates over time. The term **escalation** is usually construed to mean an increase in price over time; **de-escalation** indicates a decrease in price. In this discussion, escalation means a change in price in either direction. The question of whether or not to escalate prices in an income stream depends on the information available and the ability to translate that information into expectation.

## Crude Oil and Natural Gas as Commodities

Crude oil and natural gas are **open-market commodities**, where prices are determined by the interaction of buyers and sellers (broadly speaking) but are also subject to a wide range of economic and political/regulatory forces which can cause prices to rise and fall irrespective of the intentions of market participants. When making a projection of future prices, the commodity nature of crude oil, as well as natural gas and other products, should be borne in mind. A few major points to ponder:

- Oil is fungible: It is readily moved from place to place. Depending on the ability of refineries to process the crude, oil from one source may be substituted for oil from another source; if oil is not available from Venezuela it may be available from the Persian Gulf or through increased domestic production and/or imports from Canada via pipeline.
- Oil markets are regional or global: A price change in Qatar will affect oil prices in West Texas.

- Changes in oil prices bring about changes in inflation: Inflation does not increase oil prices. Fuel costs are part of the Consumer Price Index (CPI). Inflation does impact the cost of producing and processing oil which can have a long-term impact on oil supply.
- The price of oil is subject to two very broad influences: supply/demand and political/regulatory interference (or influence, take your pick). While the latter is often temporarily important, such as through price controls, embargos, restrictions on licensing and development, and preferences accorded other fuels, supply/demand is always dominant. Experience has also shown that when oil prices get too high, market forces act to increase competition for production that effectively caps the price and eventually causes the price to decline as additional supplies and energy sources come onstream.

## Oil Price Projection

This brings us to the next question in estimating future prices and, by extension, future income streams. Should prices be expected to remain at the "current" level or should prices be expected to change? If the latter, at what rate(s)? If prices are expected to increase (or decline), should the increase/decline be "capped" or limited at some predetermined price? Depending on the circumstances and purpose of the evaluation, the decision can have an impact on the results of the evaluation. In some circumstances, such as internal evaluations, the procedure is dictated; it may be one or the other, or both: a short-term increase followed by long-term flat or decline. In most situations there is room for judgement. However, it must be informed judgement and the evaluator must consider the commodity nature of oil and natural gas and be aware of the historic performance of product prices relative to market conditions.

There are at least two broad schools of thought on this issue. The **flat price** school opines that the initial price, however determined, should be projected unchanged (flat) over the life of the projection. Thus, if the representative initial oil price is $50 per barrel, then $50 is used throughout the evaluation of that property, with no changes. The **escalation** school suggests that initial price(s) should be increased and/or decreased over time according to perceived expectations for future oil and/or gas markets as of the time of the evaluation as well as the ability to articulate and support those expectations.

*Bearing in mind that any price projection will likely be wrong,* rational arguments can be made for either approach. The argument for flat pricing is simplicity; only the determination of a representative initial price is necessary. The initial price may be taken as an average over a representative period prior to the date of evaluation and, to the extent that the price history has been studied and incorporated into the selection of the initial price, a projection that maintains the initial price without change can be a rational approach.

The strongest argument for projecting changes in prices is the observation that prices do change. A rational appraisal should reflect real markets. The argument against – or perhaps reticence regarding – escalation has been an awareness of the experience of industry. While it is readily apparent that oil prices do not remain flat and that they do change, suggesting that some form of "escalation" may be appropriate, there are also numerous examples of price expectations that were, in retrospect, excessive. In the late 1970s and early 1980s, largely as the result of turmoil in the Middle East and production restrictions by the Organization of Petroleum Exporting Countries (OPEC), oil prices were projected to reach $100 per barrel within a few years. The drilling rig count set a record, properties were bought and sold, new technologies were developed and old ones improved to allow greater production, regulations and taxes increased, and a large amount of new production came into being based on the expectation of continually increasing oil prices. However, by 1986-87, oil prices had retreated to pre-1979 levels. Oil prices did not reach $100 per barrel until 2011.

Changes in price, either increasing or declining, have occurred with some regularity. These changes have been observed to be of short duration, often resulting in a return to a more stable level. As an illustration, the SPEE conducts an annual survey of persons who evaluate oil & gas properties as a profession. Part of the survey is a projection of future oil & gas prices as of the date of the survey. These projections, taken as an aggregation of the individual projections for each year, almost always show increasing prices, some years more aggressively than others, that are just as invariably incorrect. The point of this is not that an oil price escalation is incorrect – the escalation of the oil price in income projections which turned out to be incorrect was not the culprit, only the symptom. The problem was, and remains to some extent, a reluctance to temper expectations with realistic market analysis. In a market economy, no relation is static – all relations are dynamic. If the oil or gas price is really anticipated to increase for some reason, it must also be expected that, in a free market, there will be a reaction in the form of greater production, reduced demand, conservation, use of alternative fuels, or all of the above, which will eventually impose some restraint on price. The restraint could take the form of a slower rate of price increase and an eventual decline in price.

A rational approach to a price projection would be to determine the market level price trend over a long period (5-10 years), project the initial price to return to the market level price over a reasonable period of time, and then hold the market level price flat for the remainder of the projection. This approach avoids the "flat" price problem of projecting a short-term high or low price over the remaining life of the property and also avoids the "escalated" price tendency to assume that prices will always increase (or decrease) over long periods. Admittedly, this approach requires considerable amount of data collection and analysis but if the result is a rational and defensible projection, then the effort may be worthwhile.

## Gas Price Projection

Estimating future natural gas prices is similar to projecting oil prices, but there are important differences. The same conditions of supply and demand apply, but the market for gas is subject to two additional constraints. First, gas is not as fungible as oil; it is not as readily moved from market to market. The market is at the end of a pipeline to somewhere, making the value of the gas subject to the market at the end of the line as well as at multiple gathering system connections. This constraint has been alleviated to some extent by the use of shorter contracts and spot market sales, but markets tend to remain regional. Second, gas is a comparatively more highly regulated commodity than crude oil, with federal, state, and, in some instances, local authorities having differing levels of control over gas distribution and pricing. In this regulated environment, the best approach may be to review the price history of the property, the contract terms (if any), and, if possible, the price performance of similar properties in the same market. Depending on the purpose of the evaluation, some allowance can be made for reasonable changes in contract terms, market conditions, and existing regulatory conditions.

## Other Product Prices

Projecting prices for NGL, propane, butane, sulfur, or certain exotic products is generally more complex than for crude oil and natural gas. The primary reasons are market stability and variable processing costs. Product markets are not always consistent over time. Shipments may be seasonal or subject to transport availability, which may affect sales volumes and price. Processing costs, particularly that portion subject to inflation, may be expected to affect product price. This is an area where the evaluator must exercise some judgement in how to project produced volume, sales volumes, and product price. It may be that historical sales data are sufficient to allow a reasonable forecast to be extrapolated onto future production. In most evaluations these products have limited impact on value, but that does not reduce the requirement for diligence.

# Use of Oil & Gas Futures Markets

There is a dichotomy in this issue that must be resolved by the evaluator. While the historical perspective clearly shows that long-term price escalations do not prevail, many segments of the oil industry continue to publish studies and/or predictions of future long-term price increases. If these latter are considered to be data for evaluation purposes, how can they be reconciled with historical performance? The optimal case is to employ price projections that result from the actions of persons involved in the buying and selling of crude oil and natural gas.

Stepping back a bit, crude oil and natural gas (in this discussion all reference is to crude oil unless specific mention of gas is necessary) can have two prices at the same time – a **spot price** and a **futures price**. Both are a function of "physical" markets in that their purpose is to exchange actual volumes of oil and/or gas between buyers and sellers. The main difference between spot and futures prices are the timing of the transaction and the date of delivery. The spot price of a volume of oil is the present cost, for current purchase and delivery. The transaction occurs on the spot and reflects the price at which the oil is being traded in the marketplace.[13] The oil volumes that are bought and sold in the spot market happen promptly – the buyer accepts delivery of the goods right after the money is exchanged. The posted prices discussed above as a source for initial price information are examples of the spot market for specific fields and crude types. The price of the spot contract demonstrates the immediate or current market price.

On the other hand, the price of oil in the future may be estimated by way of a futures contract.

The recognition of oil & gas as tradable commodities has led to the development of vehicles for estimating the price that would be paid for a future delivery of a certain volume of oil and/or gas. The New York Mercantile Exchange or "NYMEX" (now CME Group) maintains markets in crude oil and natural gas by the trading of contracts for future delivery of a specific volume of crude oil or natural gas. These **futures markets** operate in the same manner as the stock exchanges; the prices for future contracts vary continuously over a trading session and from day to day. The resulting market activity is published daily by market sources, newspapers, and various on-line reporting agents. Oil contracts can be defined for up to 10 years, while natural gas contracts extend up to 12 years. The contract prices can give a sense of the direction of future prices (up or down) and the anticipated degree of change in prices.

There are two major oil futures contracts. West Texas Intermediate ("WTI") crude is the benchmark crude for North America and trades on the NYMEX.[14] Brent Crude is the benchmark for Africa, Europe, and the Middle East and trades on the Intercontinental Exchange.

The futures contract shows how much buyers are willing to pay for delivery of a certain volume of oil at a predetermined date. In the absence of otherwise reliable guidance regarding future oil & gas prices, an evaluator can make use of the futures market for crude oil and for natural gas (and other products) as a source of data. A futures contract represents trading a commodity for delivery at a later date and reflects a "future" price agreed upon by the buyer and seller for a specific volume of crude oil. An oil futures contract is a legal agreement to trade a particular volume of oil at a predetermined price at a specific date.

Oil futures contracts are simple in theory. The specifications for crude oil futures contracts are set by the exchange in a way that allows market participants to trade the contracts uniformly. Each contract covers 1,000 barrels of crude, and dates for delivery are available up to 10 years into the future. The NYMEX crude oil futures contract involves buying and selling at a specific location in the North American pipeline network. At some point during the delivery month, the seller must deliver the oil to the buyer at a pipeline or storage facility at the energy hub of Cushing, Oklahoma, with official transfer of title accompanying the physical movement of oil. The Cushing, Oklahoma location was chosen because of the amount of oil storage capacity located there and the interconnection to pipelines serving virtually all of the United States.

Futures contracts differ from spot or forward contracts in important ways:

1. Futures contracts are highly standardized and tightly defined in terms of the quantity and quality of oil that makes up a single contract, the delivery location, and date of delivery.
2. Futures contracts are traded through financial exchanges instead of directly between buyers and sellers.
3. Futures contracts are "financial" where the contract is settled in cash instead of through delivery of the commodity.

Companies and investors who buy in the futures market do so to lock in a price for crude oil that will be needed in the future.

> As an example, a refining company may buy crude oil contracts covering several months or years in the future to be assured of a reliable volume of crude and avoid having to deal with volatile spot prices whenever crude is needed.[15]

The spot and futures oil prices are both indicators of the overall state of the market as well as expectations for the market for that commodity. If the futures prices are less compared to the spot prices, this means that buyers are expecting the market for crude oil to decline. On the other hand, if buyers are willing to pay a premium over the spot price, this means they are anticipating the price of crude to increase.

## Application of Futures Market Data

NYMEX (CME) provides the platform for trading in crude oil contracts from one month in advance up to 10 years forward. Information regarding crude oil and natural gas futures contracts which can be used to estimate future prices for property evaluation is available from a large number of sources including the NYMEX (CME) exchange as well as online sites. Figure 36 shows a portion of the market for crude oil futures contracts as of the close of trading on February 18, 2025 as reported by *Barchart.com.* The contract data available from Barchart is through January 2036 but is truncated to December 2027 for discussion purposes. [Note: Barchart offers pricing information on a continuing active basis throughout the day in the same manner as the stock exchanges.] The data in the example is for monthly contracts for 1,000 barrels of WTI (or equivalent) crude oil to be delivered in March 2025 through December 2027. The example contains a lot of information. The table shows: the Contract Month (date of delivery), the Last quoted price of the contract crude, the change in price from the Previous closing price, the previous Opening, Highest, and Lowest price, the price Two Days previous, the previous day Trading Volume, and the Open Interest. The table also shows the price for February 2025, which is for contracts that have expired at the end of January 2025 – effectively the spot price.

**Crude Oil WTI Apr '25 (CLJ25)**
**72.28 +0.45 (+0.63%)**
72.27 x 28 72.29 x 23

| Contract | Last | Change | Open | High | Low | Previous | Volume | Open Interest | Time |
|---|---|---|---|---|---|---|---|---|---|
| + CLY00 (Cash) | 71.86 | +1.16 | 0.00 | 71.86 | 71.86 | 70.70 | 0 | 0 | 02/18/25 |
| + CLH25 (Mar '25) | 71.85 | +1.11 | 70.70 | 72.07 | 70.12 | 70.74 | 99,343 | 81,052 | 02/18/25 |
| + CLJ25 (Apr '25) | 71.83 | +1.12 | 70.69 | 72.04 | 70.12 | 70.71 | 308,150 | 289,399 | 02/18/25 |
| + CLK25 (May '25) | 71.58 | +1.02 | 70.48 | 71.79 | 70.00 | 70.56 | 111,449 | 149,432 | 02/18/25 |
| + CLM25 (Jun '25) | 71.26 | +0.92 | 70.28 | 71.47 | 69.79 | 70.34 | 94,757 | 174,443 | 02/18/25 |
| + CLN25 (Jul '25) | 70.87 | +0.84 | 69.90 | 71.09 | 69.51 | 70.03 | 36,015 | 88,998 | 02/18/25 |
| + CLQ25 (Aug '25) | 70.43 | +0.78 | 69.41 | 70.64 | 69.15 | 69.65 | 24,645 | 57,792 | 02/18/25 |
| + CLU25 (Sep '25) | 69.97 | +0.72 | 68.97 | 70.17 | 68.78 | 69.25 | 42,621 | 103,482 | 02/18/25 |
| + CLV25 (Oct '25) | 69.50 | +0.65 | 68.38 | 69.71 | 68.38 | 68.85 | 15,124 | 61,172 | 02/18/25 |
| + CLX25 (Nov '25) | 69.08 | +0.59 | 68.47 | 69.30 | 68.47 | 68.49 | 11,681 | 46,184 | 02/18/25 |
| + CLZ25 (Dec '25) | 68.70 | +0.55 | 67.86 | 68.94 | 67.65 | 68.15 | 61,375 | 182,523 | 02/18/25 |
| + CLF26 (Jan '26) | 68.35 | +0.52 | 67.80 | 68.43 | 67.80 | 67.83 | 2,302 | 41,709 | 02/18/25 |
| + CLG26 (Feb '26) | 68.04 | +0.50 | 67.92 | 68.11 | 67.83 | 67.54 | 1,764 | 27,029 | 02/18/25 |
| + CLH26 (Mar '26) | 67.75 | +0.46 | 67.49 | 67.99 | 67.19 | 67.29 | 6,783 | 45,969 | 02/18/25 |
| + CLJ26 (Apr '26) | 67.52 | +0.44 | 67.88 | 67.88 | 67.48 | 67.08 | 450 | 16,488 | 02/18/25 |
| + CLK26 (May '26) | 67.32 | +0.43 | 0.00 | 67.32 | 67.32 | 66.89 | 283 | 14,630 | 02/18/25 |
| + CLM26 (Jun '26) | 67.13 | +0.41 | 66.45 | 67.38 | 66.24 | 66.72 | 14,079 | 94,838 | 02/18/25 |
| + CLN26 (Jul '26) | 66.91 | +0.39 | 0.00 | 66.91 | 66.91 | 66.52 | 282 | 15,413 | 02/18/25 |
| + CLQ26 (Aug '26) | 66.70 | +0.38 | 0.00 | 66.70 | 66.70 | 66.32 | 468 | 10,438 | 02/18/25 |
| + CLU26 (Sep '26) | 66.51 | +0.36 | 0.00 | 66.59 | 65.99 | 66.15 | 862 | 20,175 | 02/18/25 |
| + CLV26 (Oct '26) | 66.34 | +0.34 | 0.00 | 66.34 | 66.34 | 66.00 | 223 | 12,545 | 02/18/25 |
| + CLX26 (Nov '26) | 66.19 | +0.33 | 0.00 | 66.19 | 66.19 | 65.86 | 238 | 11,957 | 02/18/25 |
| + CLZ26 (Dec '26) | 66.05 | +0.32 | 65.42 | 66.34 | 65.29 | 65.73 | 20,911 | 81,043 | 02/18/25 |
| + CLF27 (Jan '27) | 65.87 | +0.31 | 0.00 | 65.87 | 65.87 | 65.56 | 172 | 5,724 | 02/18/25 |
| + CLG27 (Feb '27) | 65.69 | +0.29 | 65.77 | 65.77 | 65.69 | 65.40 | 133 | 2,224 | 02/18/25 |
| + CLH27 (Mar '27) | 65.53 | +0.28 | 65.76 | 65.77 | 65.53 | 65.25 | 423 | 7,682 | 02/18/25 |
| + CLJ27 (Apr '27) | 65.39 | +0.27 | 65.66 | 65.66 | 65.39 | 65.12 | 98 | 1,172 | 02/18/25 |
| + CLK27 (May '27) | 65.28 | +0.24 | 0.00 | 65.28 | 65.28 | 65.04 | 32 | 1,974 | 02/18/25 |
| + CLM27 (Jun '27) | 65.17 | +0.24 | 64.92 | 65.50 | 64.92 | 64.93 | 2,773 | 21,909 | 02/18/25 |
| + CLN27 (Jul '27) | 65.02 | +0.23 | 0.00 | 65.02 | 65.02 | 64.79 | 26 | 1,854 | 02/18/25 |
| + CLQ27 (Aug '27) | 64.88 | +0.22 | 64.96 | 65.01 | 64.88 | 64.66 | 3 | 814 | 02/18/25 |
| + CLU27 (Sep '27) | 64.78 | +0.21 | 65.10 | 65.11 | 64.78 | 64.57 | 43 | 5,135 | 02/18/25 |
| + CLV27 (Oct '27) | 64.69 | +0.20 | 0.00 | 64.69 | 64.69 | 64.49 | 0 | 811 | 02/18/25 |
| + CLX27 (Nov '27) | 64.61 | +0.18 | 0.00 | 64.79 | 64.61 | 64.43 | 0 | 2,110 | 02/18/25 |
| + CLZ27 (Dec '27) | 64.52 | +0.16 | 64.33 | 64.97 | 64.01 | 64.36 | 4,878 | 48,740 | 02/18/25 |

**Figure 36.** Crude Oil WTI Prices and Crude Oil WTI Futures Prices (Barchart.com)

The most useful information for evaluation purposes is the Last Price, the Volume (of trades), and the Open Interest (the number of outstanding contracts). The Last (current) price for a contract for March 2025 is $71.85 per barrel; the Last price for crude to be delivered in December 2025 is $74.68 while the price for delivery in December 2027 is $61.88 per barrel. The times series of futures prices on a given date is called the **forward curve** and represents the best expectations of the market on a specific date as to the direction of the price of oil.

The time series of the futures contract price is for WTI crude. Unless the subject property actually produces WTI oil, the projection is not directly applicable. However, the WTI futures market is a proxy for the market for all crude oil in North America. The projection of Last Prices can be used by calculating the change in prices from contract month to month (Exhibit 3) and then applying the *trend* of WTI price changes to the initial oil price for the subject property. As an historical example, assume that the initial price for Kern Front crude oil is $64.57 as of December 30, 2021. A projection of that price using the NYMEX price trend through December 2031 is presented in Exhibit 3 and Figure 37. This approach works well because it is derived from readily available public data and can be calculated for different starting dates.

There are a couple of caveats to be considered.

- First, it is the futures price *trend line* of derived changes in monthly prices, not the actual WTI contract price, that is to be used.
- Second, Figure 36 shows that the trading volume (number of contract trades) is not uniform and becomes rather sparse after 2-3 years, suggesting that the trend line after the first few years may be interpolated and, while still useful, should be used with caution. It is notable that in later years such as 2025 and 2026 there are a much larger number of outstanding contracts for June and particularly December than in other months.
- Third, there may be differences between contract crude and the oil on the subject property that are not necessarily reflected in the NYMEX market, which may require adjustments to the projection.
- Lastly, the futures contract market may not always be consistent or uniform over the entire period of coverage; that is, there may be periods when there are few contracts being traded or contracts may be being traded for a specific purpose which may result in prices that are not consistent with prices prior to or subsequent to the trading period.

**Price Projections for Kern County Properties**
**Initial Price is Average of Six Months Prior**
**NYMEX Crude Oil Contract Prices**
**December 20, 2021**

| Contract Date | Settle Price $/Bbl | Price Change % | Price Projection $/Bbl | Yearly Average Price |
|---|---|---|---|---|
| Nov-21 | 79.18 | | 72.15 | |
| Cash | 70.86 | 0.89492 | 64.57 | |
| Jan-22 | 66.77 | 0.94228 | 60.84 | |
| Feb-22 | 67.17 | 1.00599 | 61.21 | |
| Mar-22 | 66.92 | 0.99628 | 60.98 | |
| Apr-22 | 66.65 | 0.99597 | 60.73 | |
| May-22 | 66.30 | 0.99475 | 60.41 | |
| Jun-22 | 65.99 | 0.99532 | 60.13 | |
| Jul-22 | 65.67 | 0.99515 | 59.84 | |
| Aug-22 | 65.29 | 0.99421 | 59.49 | |
| Sep-22 | 64.90 | 0.99403 | 59.14 | |
| Oct-22 | 64.56 | 0.99476 | 58.83 | |
| Nov-22 | 63.59 | 0.98498 | 57.94 | |
| Dec-22 | 63.86 | 1.00425 | 58.19 | 59.811 |
| Jan-23 | 63.65 | 0.99671 | 58.00 | |
| Feb-23 | 65.61 | 1.03079 | 59.78 | |
| Mar-23 | 65.24 | 0.99436 | 59.45 | |
| Apr-23 | 62.91 | 0.96429 | 57.32 | |
| May-23 | 64.60 | 1.02686 | 58.86 | |
| Jun-23 | 61.92 | 0.95851 | 56.42 | |
| Jul-23 | 63.99 | 1.03343 | 58.31 | |
| Aug-23 | 63.69 | 0.99531 | 58.04 | |
| Sep-23 | 63.42 | 0.99576 | 57.79 | |
| Oct-23 | 63.19 | 0.99637 | 57.58 | |
| Nov-23 | 62.99 | 0.99683 | 57.40 | |
| Dec-23 | 61.00 | 0.96841 | 55.58 | 57.878 |
| Jan-24 | 60.35 | 0.98934 | 54.99 | |
| Feb-24 | 62.32 | 1.03264 | 56.79 | |
| Mar-24 | 62.08 | 0.99615 | 56.57 | |
| Apr-24 | 61.87 | 0.99662 | 56.38 | |
| May-24 | 61.70 | 0.99725 | 56.22 | |
| Jun-24 | 59.99 | 0.97229 | 54.66 | |
| Jul-24 | 61.32 | 1.02217 | 55.88 | |
| Aug-24 | 61.13 | 0.99690 | 55.70 | |
| Sep-24 | 60.95 | 0.99706 | 55.54 | |
| Oct-24 | 60.80 | 0.99754 | 55.40 | |
| Nov-24 | 59.48 | 0.97829 | 54.20 | |
| Dec-24 | 59.20 | 0.99529 | 53.94 | 55.523 |
| Jan-25 | 60.34 | 1.01926 | 54.98 | |
| Feb-25 | 60.17 | 0.99718 | 54.83 | |
| Mar-25 | 60.02 | 0.99751 | 54.69 | |
| Apr-25 | 59.89 | 0.99783 | 54.57 | |
| May-25 | 59.78 | 0.99816 | 54.47 | |
| Jun-25 | 59.00 | 0.98695 | 53.76 | |
| Jul-25 | 59.47 | 1.00797 | 54.19 | |
| Aug-25 | 59.33 | 0.99765 | 54.06 | |
| Sep-25 | 59.21 | 0.99798 | 53.95 | |
| Oct-25 | 59.12 | 0.99848 | 53.87 | |
| Nov-25 | 59.00 | 0.99797 | 53.76 | |
| Dec-25 | 57.74 | 0.97864 | 52.61 | 54.147 |
| Jan-26 | 58.77 | 1.01784 | 53.55 | |
| Feb-26 | 58.63 | 0.99762 | 53.42 | |
| Mar-26 | 58.51 | 0.99795 | 53.32 | |
| Apr-26 | 58.41 | 0.99829 | 53.22 | |
| May-26 | 58.36 | 0.99914 | 53.18 | |
| Jun-26 | 58.30 | 0.99897 | 53.12 | |
| Jul-26 | 58.21 | 0.99846 | 53.04 | |
| Aug-26 | 58.06 | 0.99742 | 52.91 | |
| Sep-26 | 58.00 | 0.99897 | 52.85 | |
| Oct-26 | 57.96 | 0.99931 | 52.81 | |
| Nov-26 | 57.93 | 0.99948 | 52.79 | |
| Dec-26 | 57.90 | 0.99948 | 52.76 | 53.081 |

**Price Projections for Kern County Properties**
**Initial Price is Average of Six Months Prior**
**NYMEX Crude Oil Contract Prices**
**December 20, 2021**

| Contract Date | Settle Price $/Bbl | Price Change % | Price Projection $/Bbl | Yearly Average Price |
|---|---|---|---|---|
| Jan-27 | 57.82 | 0.99862 | 52.69 | |
| Feb-27 | 57.70 | 0.99792 | 52.58 | |
| Mar-27 | 57.60 | 0.99827 | 52.49 | |
| Apr-27 | 57.53 | 0.99878 | 52.42 | |
| May-27 | 57.54 | 1.00017 | 52.43 | |
| Jun-27 | 57.53 | 0.99983 | 52.42 | |
| Jul-27 | 57.48 | 0.99913 | 52.38 | |
| Aug-27 | 57.39 | 0.99843 | 52.29 | |
| Sep-27 | 57.34 | 0.99913 | 52.25 | |
| Oct-27 | 57.32 | 0.99965 | 52.23 | |
| Nov-27 | 57.35 | 1.00052 | 52.26 | |
| Dec-27 | 57.39 | 1.00070 | 52.29 | 52.394 |
| Jan-28 | 57.35 | 0.99930 | 52.26 | |
| Feb-28 | 57.37 | 1.00035 | 52.28 | |
| Mar-28 | 57.33 | 0.99930 | 52.24 | |
| Apr-28 | 57.37 | 1.00070 | 52.28 | |
| May-28 | 57.36 | 0.99983 | 52.27 | |
| Jun-28 | 57.33 | 0.99948 | 52.24 | |
| Jul-28 | 57.41 | 1.00140 | 52.31 | |
| Aug-28 | 57.41 | 1.00000 | 52.31 | |
| Sep-28 | 57.38 | 0.99948 | 52.29 | |
| Oct-28 | 57.39 | 1.00017 | 52.29 | |
| Nov-28 | 57.39 | 1.00000 | 52.29 | |
| Dec-28 | 57.32 | 0.99878 | 52.23 | 52.274 |
| Jan-29 | 57.40 | 1.00140 | 52.30 | |
| Feb-29 | 57.39 | 0.99983 | 52.29 | |
| Mar-29 | 57.43 | 1.00070 | 52.33 | |
| Apr-29 | 57.45 | 1.00035 | 52.35 | |
| May-29 | 57.45 | 1.00000 | 52.35 | |
| Jun-29 | 57.28 | 0.99704 | 52.19 | |
| Jul-29 | 57.30 | 1.00035 | 52.21 | |
| Aug-29 | 57.29 | 0.99983 | 52.20 | |
| Sep-29 | 57.28 | 0.99983 | 52.19 | |
| Oct-29 | 57.26 | 0.99965 | 52.18 | |
| Nov-29 | 57.23 | 0.99948 | 52.15 | |
| Dec-29 | 57.20 | 0.99948 | 52.12 | 52.240 |
| Jan-30 | 57.24 | 1.00070 | 52.16 | |
| Feb-30 | 57.27 | 1.00052 | 52.19 | |
| Mar-30 | 57.27 | 1.00000 | 52.19 | |
| Apr-30 | 57.27 | 1.00000 | 52.19 | |
| May-30 | 57.27 | 1.00000 | 52.19 | |
| Jun-30 | 57.28 | 1.00017 | 52.19 | |
| Jul-30 | 57.32 | 1.00070 | 52.23 | |
| Aug-30 | 57.37 | 1.00087 | 52.28 | |
| Sep-30 | 57.39 | 1.00035 | 52.29 | |
| Oct-30 | 57.42 | 1.00052 | 52.32 | |
| Nov-30 | 57.44 | 1.00035 | 52.34 | |
| Dec-30 | 57.45 | 1.00017 | 52.35 | 52.242 |
| Jan-31 | 47.74 | 0.83098 | 43.50 | |
| Feb-31 | 53.24 | 1.11521 | 48.51 | |
| Mar-31 | 61.49 | 1.15496 | 56.03 | |
| Apr-31 | 61.55 | 1.00098 | 56.09 | |
| May-31 | 62.44 | 1.01446 | 56.90 | |
| Jun-31 | 62.05 | 0.99375 | 56.54 | |
| Jul-31 | 73.06 | 1.17744 | 66.57 | |
| Aug-31 | 67.42 | 0.92280 | 61.43 | |
| Sep-31 | 62.32 | 0.92435 | 56.79 | |
| Oct-31 | 70.56 | 1.13222 | 64.30 | |
| Nov-31 | 83.87 | 1.18863 | 76.42 | |
| Dec-31 | 76.10 | 0.90736 | 69.34 | 59.369 |

**Exhibit 3.** Kern County Properties price projections

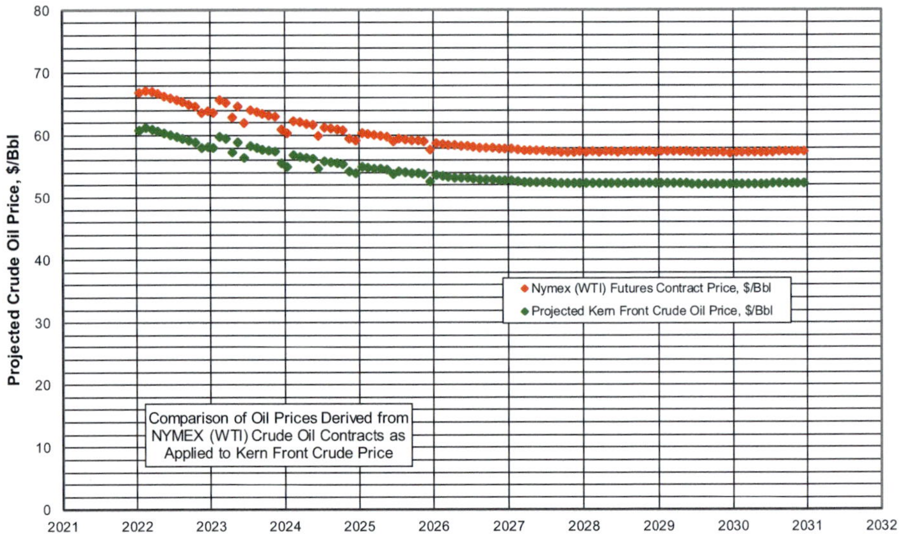


**Figure 37.** Use of NYMEX Crude Oil Futures Contract Prices to Project Oil Prices for Property Evaluation

The NYMEX supports a futures market for natural gas that operates in the same manner as the market for crude oil futures and can be used in the same manner to estimate future prices. There are a couple of minor differences. Gas contracts are based on delivery at a central location known as the Henry Hub; prices are shown in Exhibit 4 for an effective date of January 1, 2023. In this case, the NYMEX projects for two dates on either side of the effective date (December 30, 2022 and January 3, 2023) were averaged to provide a representative projection.

There is also a market for futures contracts for other products such as Gasoline, Butane, and Propane that could be used as a proxy for projection of NGL prices.

# Deductions from Crude Oil Prices and Income

Deductions from product prices are charges against the product sales income that could be treated as operating costs in an evaluation. However, they are discussed here because they are the types of costs that, in some cases, are deducted by the crude oil or natural gas purchaser before payment is made for the oil or gas. In those circumstances, an evaluator has the option of using the oil or gas price before deductions and then including the charges as operating costs, or using the "net" price after taking the deductions. In some cases, the charges or deductions are derived from sales contracts. Defining a history of deductions is another good argument for analyzing crude oil or natural gas sales statements for the property being evaluated. If deductions for transportation, etc., are being made by the purchaser, a sales history of the deductions relative to sales volume over several months or years, if available, would define the nature and provide the amounts of deductions

**Marcellus Natural Gas**
**NYMEX Price December 30, 2022**
**Starting price is average of July through December, 2022**

| Contract Month | Settle Price $/MCF | Difference | Actual Price $/MCF | Yearly Average Price |
|---|---|---|---|---|
| Dec-22 | 5.770 | | 5.798 | |
| Cash Price | 4.559 | 0.79012 | 4.581 | |
| Feb-23 | 4.559 | 1.00000 | 4.581 | |
| Mar-23 | 4.117 | 0.90305 | 4.137 | |
| Apr-23 | 3.925 | 0.95336 | 3.944 | |
| May-23 | 3.934 | 1.00229 | 3.953 | |
| Jun-23 | 4.053 | 1.03025 | 4.073 | |
| Jul-23 | 4.168 | 1.02837 | 4.188 | |
| Aug-23 | 4.171 | 1.00072 | 4.191 | |
| Sep-23 | 4.114 | 0.98633 | 4.134 | |
| Oct-23 | 4.173 | 1.01434 | 4.193 | |
| Nov-23 | 4.477 | 1.07285 | 4.499 | |
| Dec-23 | 4.873 | 1.08845 | 4.897 | 4.281 |
| Jan-24 | 5.117 | 1.05007 | 5.142 | |
| Feb-24 | 4.932 | 0.96385 | 4.956 | |
| Mar-24 | 4.387 | 0.88950 | 4.408 | |
| Apr-24 | 3.835 | 0.87417 | 3.854 | |
| May-24 | 3.793 | 0.98905 | 3.811 | |
| Jun-24 | 3.877 | 1.02215 | 3.896 | |
| Jul-24 | 3.969 | 1.02373 | 3.988 | |
| Aug-24 | 3.994 | 1.00630 | 4.013 | |
| Sep-24 | 3.955 | 0.99024 | 3.974 | |
| Oct-24 | 4.043 | 1.02225 | 4.063 | |
| Nov-24 | 4.369 | 1.08063 | 4.390 | |
| Dec-24 | 4.856 | 1.11147 | 4.880 | 4.281 |
| Jan-25 | 5.157 | 1.06199 | 5.182 | |
| Feb-25 | 4.970 | 0.96374 | 4.994 | |
| Mar-25 | 4.550 | 0.91549 | 4.572 | |
| Apr-25 | 3.960 | 0.87033 | 3.979 | |
| May-25 | 3.921 | 0.99015 | 3.940 | |
| Jun-25 | 4.011 | 1.02295 | 4.030 | |
| Jul-25 | 4.096 | 1.02119 | 4.116 | |
| Aug-25 | 4.136 | 1.00977 | 4.156 | |
| Sep-25 | 4.126 | 0.99758 | 4.146 | |
| Oct-25 | 4.225 | 1.02399 | 4.246 | |
| Nov-25 | 4.548 | 1.07645 | 4.570 | |
| Dec-25 | 4.983 | 1.09565 | 5.007 | 4.412 |
| Jan-26 | 5.277 | 1.05900 | 5.303 | |
| Feb-26 | 5.062 | 0.95926 | 5.087 | |
| Mar-26 | 4.641 | 0.91683 | 4.664 | |
| Apr-26 | 4.031 | 0.86856 | 4.051 | |
| May-26 | 3.979 | 0.98710 | 3.998 | |
| Jun-26 | 4.058 | 1.01985 | 4.078 | |
| Jul-26 | 4.143 | 1.02095 | 4.163 | |
| Aug-26 | 4.173 | 1.00724 | 4.193 | |
| Sep-26 | 4.163 | 0.99760 | 4.183 | |
| Oct-26 | 4.258 | 1.02282 | 4.279 | |
| Nov-26 | 4.585 | 1.07680 | 4.607 | |
| Dec-26 | 5.025 | 1.09597 | 5.049 | 4.471 |
| Jan-27 | 5.285 | 1.05174 | 5.311 | |
| Feb-27 | 5.065 | 0.95837 | 5.090 | |
| Mar-27 | 4.675 | 0.92300 | 4.698 | |
| Apr-27 | 4.060 | 0.86845 | 4.080 | |
| May-27 | 4.034 | 0.99360 | 4.054 | |
| Jun-27 | 4.115 | 1.02008 | 4.135 | |
| Jul-27 | 4.198 | 1.02017 | 4.218 | |
| Aug-27 | 4.231 | 1.00786 | 4.252 | |
| Sep-27 | 4.230 | 0.99976 | 4.251 | |
| Oct-27 | 4.313 | 1.01962 | 4.334 | |
| Nov-27 | 4.648 | 1.07767 | 4.671 | |
| Dec-27 | 5.073 | 1.09144 | 5.098 | 4.516 |
| Jan-28 | 5.333 | 1.05125 | 5.359 | |
| Feb-28 | 5.113 | 0.95875 | 5.138 | |
| Mar-28 | 4.738 | 0.92666 | 4.761 | |
| Apr-28 | 4.118 | 0.86914 | 4.138 | |
| May-28 | 4.110 | 0.99806 | 4.130 | |
| Jun-28 | 4.194 | 1.02044 | 4.214 | |
| Jul-28 | 4.281 | 1.02074 | 4.302 | |
| Aug-28 | 4.305 | 1.00561 | 4.326 | |
| Sep-28 | 4.305 | 1.00000 | 4.326 | |
| Oct-28 | 4.388 | 1.01928 | 4.409 | |
| Nov-28 | 4.737 | 1.07954 | 4.760 | |
| Dec-28 | 5.161 | 1.08951 | 5.186 | 4.587 |

**Marcellus Natural Gas**
**NYMEX Price January 3, 2023**
**Starting price is average of July through December, 2022**

| Contract Month | Settle Price $/MCF | Difference | Actual Price $/MCF | Yearly Average Price | Price | |
|---|---|---|---|---|---|---|
| Dec-22 | 5.770 | | 5.798 | | 5.798 | |
| Cash Price | 4.475 | 0.77556 | 4.497 | | 4.539 | |
| Feb-23 | 3.951 | 0.88291 | 3.970 | | 4.276 | |
| Mar-23 | 3.675 | 0.93014 | 3.693 | | 3.915 | |
| Apr-23 | 3.583 | 0.97497 | 3.600 | | 3.772 | |
| May-23 | 3.609 | 1.00726 | 3.627 | | 3.790 | |
| Jun-23 | 3.738 | 1.03574 | 3.756 | | 3.914 | |
| Jul-23 | 3.858 | 1.03210 | 3.877 | | 4.032 | |
| Aug-23 | 3.876 | 1.00467 | 3.895 | | 4.043 | |
| Sep-23 | 3.826 | 0.98710 | 3.845 | | 3.989 | |
| Oct-23 | 3.893 | 1.01751 | 3.912 | | 4.053 | |
| Nov-23 | 4.252 | 1.09222 | 4.273 | | 4.386 | |
| Dec-23 | 4.689 | 1.10278 | 4.712 | 3.971 | 4.804 | 4.126 |
| Jan-24 | 4.947 | 1.05502 | 4.971 | | 5.056 | |
| Feb-24 | 4.774 | 0.96503 | 4.797 | | 4.877 | |
| Mar-24 | 4.256 | 0.89150 | 4.277 | | 4.342 | |
| Apr-24 | 3.726 | 0.87547 | 3.744 | | 3.799 | |
| May-24 | 3.692 | 0.99087 | 3.710 | | 3.761 | |
| Jun-24 | 3.786 | 1.02546 | 3.804 | | 3.850 | |
| Jul-24 | 3.872 | 1.02272 | 3.891 | | 3.940 | |
| Aug-24 | 3.920 | 1.01240 | 3.939 | | 3.976 | |
| Sep-24 | 3.892 | 0.99286 | 3.911 | | 3.943 | |
| Oct-24 | 3.963 | 1.01824 | 3.982 | | 4.022 | |
| Nov-24 | 4.296 | 1.08403 | 4.317 | | 4.354 | |
| Dec-24 | 4.825 | 1.12314 | 4.848 | 4.183 | 4.864 | 4.232 |
| Jan-25 | 5.103 | 1.05762 | 5.128 | | 5.155 | |
| Feb-25 | 4.983 | 0.97648 | 5.007 | | 5.001 | |
| Mar-25 | 4.505 | 0.90407 | 4.527 | | 4.549 | |
| Apr-25 | 3.905 | 0.86681 | 3.924 | | 3.952 | |
| May-25 | 3.919 | 1.00359 | 3.938 | | 3.939 | |
| Jun-25 | 4.010 | 1.02322 | 4.029 | | 4.030 | |
| Jul-25 | 4.096 | 1.02145 | 4.116 | | 4.116 | |
| Aug-25 | 4.173 | 1.01880 | 4.193 | | 4.175 | |
| Sep-25 | 4.110 | 0.98490 | 4.130 | | 4.138 | |
| Oct-25 | 4.227 | 1.02847 | 4.248 | | 4.247 | |
| Nov-25 | 4.551 | 1.07665 | 4.573 | | 4.572 | |
| Dec-25 | 4.987 | 1.09580 | 5.011 | 4.402 | 5.009 | 4.407 |
| Jan-26 | 5.287 | 1.06016 | 5.313 | | 5.308 | |
| Feb-26 | 5.077 | 0.96028 | 5.102 | | 5.094 | |
| Mar-26 | 4.656 | 0.91708 | 4.679 | | 4.671 | |
| Apr-26 | 4.046 | 0.86899 | 4.066 | | 4.058 | |
| May-26 | 3.994 | 0.98715 | 4.013 | | 4.006 | |
| Jun-26 | 4.073 | 1.01978 | 4.093 | | 4.085 | |
| Jul-26 | 4.158 | 1.02087 | 4.178 | | 4.171 | |
| Aug-26 | 4.188 | 1.00722 | 4.208 | | 4.201 | |
| Sep-26 | 4.178 | 0.99761 | 4.198 | | 4.191 | |
| Oct-26 | 4.273 | 1.02274 | 4.294 | | 4.286 | |
| Nov-26 | 4.600 | 1.07653 | 4.622 | | 4.615 | |
| Dec-26 | 5.040 | 1.09565 | 5.064 | 4.486 | 5.057 | 4.479 |
| Jan-27 | 5.300 | 1.05159 | 5.326 | | 5.318 | |
| Feb-27 | 5.080 | 0.95849 | 5.105 | | 5.097 | |
| Mar-27 | 4.684 | 0.92205 | 4.707 | | 4.702 | |
| Apr-27 | 4.096 | 0.87447 | 4.116 | | 4.098 | |
| May-27 | 4.043 | 0.98706 | 4.063 | | 4.058 | |
| Jun-27 | 4.124 | 1.02003 | 4.144 | | 4.139 | |
| Jul-27 | 4.207 | 1.02013 | 4.227 | | 4.223 | |
| Aug-27 | 4.240 | 1.00784 | 4.261 | | 4.256 | |
| Sep-27 | 4.239 | 0.99976 | 4.260 | | 4.255 | |
| Oct-27 | 4.322 | 1.01958 | 4.343 | | 4.338 | |
| Nov-27 | 4.657 | 1.07751 | 4.680 | | 4.675 | |
| Dec-27 | 5.082 | 1.09126 | 5.107 | 4.528 | 5.102 | 4.522 |
| Jan-28 | 5.342 | 1.05116 | 5.368 | | 5.363 | |
| Feb-28 | 5.122 | 0.95882 | 5.147 | | 5.142 | |
| Mar-28 | 4.747 | 0.92679 | 4.770 | | 4.766 | |
| Apr-28 | 4.127 | 0.86939 | 4.147 | | 4.143 | |
| May-28 | 4.119 | 0.99806 | 4.139 | | 4.134 | |
| Jun-28 | 4.203 | 1.02039 | 4.223 | | 4.219 | |
| Jul-28 | 4.290 | 1.02070 | 4.311 | | 4.306 | |
| Aug-28 | 4.314 | 1.00559 | 4.335 | | 4.330 | |
| Sep-28 | 4.314 | 1.00000 | 4.335 | | 4.330 | |
| Oct-28 | 4.397 | 1.01924 | 4.418 | | 4.414 | |
| Nov-28 | 4.746 | 1.07937 | 4.769 | | 4.765 | |
| Dec-28 | 5.170 | 1.08934 | 5.195 | 4.596 | 5.191 | 4.592 |

**Exhibit 4.** Marcellus Natural Gas Nymex Prices

## Transportation Costs

Transportation costs can include trucking, gathering system access, or pumping charges and can range from a few cents up to several dollars per barrel.

> As an example, production in some parts of more remote oil fields such as Cat Canyon in Santa Barbara County or San Ardo in Monterey County do not have direct access to an oil pipeline. The oil must be trucked to a refinery, pipeline, or railroad transfer point. These charges are usually negotiated at a fixed rate per barrel. As a consequence, the impact is greater when oil prices decline.

## Dehydration

Generally, crude oil must meet certain standards of water and entrained solids content, also known as basic sediment and water (BS&W), to be accepted for sale. Sediment is commonly removed or reduced through use of settling tanks and separators on the property and is an operating cost. Removal of water through dehydration is also the responsibility of the crude seller and is an operating cost. It is not unknown, however, for additional dehydration charges to be imposed by the crude purchaser to meet pipeline and/or refining specifications; this may include passing crude through oil-water separators/filters. This cost, where it occurs, is usually reflected as a reduction in the price paid for the treated oil.

## Heating

Heating of crude oil in transit through gathering systems and pipelines is often necessary when heavy gravity/high viscosity crude oil is being transported; the Trans-Alaska Pipeline is such a case. Another deduction might be for heating pipelines or transportation facilities to handle certain types of low gravity or high wax crude oil. Where these conditions are prevalent, the resulting costs may or may not be factored into the posted price either as a price, reduction or a specific deduction from the price or sales revenue. The costs of heating can be charged back to crude shippers-sellers and may be shown as deductions from the sales price. An evaluation of the property must determine whether these costs, if incurred, should be recognized as operating costs or as deductions from sales revenue.

## Diluent Separation

The production and transportation efficiency of low-gravity, high(er)-viscosity "native" oil is often improved by the mixing of lighter-gravity oil or other refinery products, such as kerosene, to serve as a diluent or solvent. The diluent can be introduced either in the wellbore through tubing to improve pumping efficiency, or it can be added in surface tanks prior to being shipped, or both. The diluent may have the practical effect of reducing the viscosity of the oil, making the mixture easier to be lifted by down-hole pumps and/or transported through gathering systems and pipelines. The admixture of lighter-gravity oil to the heavier native crude also results in a volume of modestly higher-gravity crude oil that may command a higher price than would the native crude.

In common practice, the producer obtains the diluent from another producer with higher-gravity oil production or from a local refiner. Diluent is stored in tanks on the property until injected into the producing wells and/or blended on the surface. The mixture of native crude and diluent is shipped to the crude buyer. If the producer buys the diluent for blending and then sells the blended crude for a higher-gravity price, the diluent purchase becomes an operating cost. If the crude buyer is also the seller of the diluent, the cost of the diluent may be netted out of the final price paid for the improved crude. There may also be a charge for separation and the analysis to determine the "native gravity." The producer is then paid for the native gravity oil. The sales statement will normally reflect the calculation.

> As an example, much of the crude oil produced in central California oil fields has a native gravity of 10-13° API. This crude is commonly mixed with lighter crude or refinery products (kerosene, etc.) to improve the viscosity and, marginally, oil gravity. Assume every 10 barrels of produced 11° API crude requires mixing with 2 barrels of diluent; if the blended crude (12 barrels) sells for $50 per barrel and the diluent is $55 per barrel, the native gravity value is $49 per barrel.

# Deductions from Natural Gas Prices and Income

Natural gas sales prices and income may be subject to a wide range of deductions, charges, and fees for compression, transportation, treating, dehydration, processing, marketing, and other categories. While many of these charges are a function of producing the gas (compression) and making the gas saleable at pipeline specifications (treating, dehydration, processing) and should be considered to be operating costs and, therefore, the responsibility of the operator, it is not uncommon for such charges to be imposed by gas purchasers resulting in reduced prices and income. These charges are often fixed amounts.

Recent experience in Marcellus Shale gas operations has demonstrated that there can be multiple deductions from gas sales. These deductions are usually reported as being charges by third-party operators of gathering systems and pipelines rather than by the gas producer. However they may occur, these charges must be individually defined and quantified. They can then be treated as deductions from the gas sales income or as operating costs. In either case, the charges result in a reduction in income.

An evaluation of royalty interests in a number of properties in north-central Pennsylvania found that four charges were being deducted from the natural gas sales. The charges were made by the gas collection operator after the sale of the gas to the regional pipeline. Information for the charges was obtained from the royalty statements. Average deductions over a six-month period ending with December 2022 are as follows:

| | |
|---|---|
| Gathering | $0.1990/ MCF |
| Transportation | $0.1693/ MCF |
| Dehydration | $0.0043/ MCF |
| Compression | $0.1081/ MCF |
| Total | $0.4807/MCF |

The amount of the charges varies marginally from month to month. Considering that the average gas sales price over the same period before deductions was $3.89 per MCF, the charges equate to 12.37% of the sales price. The deductions could have an impact on sales revenue particularly if gas prices were to decline. The decision whether the charges should be treated as operating costs or as reductions in the gas price is a function of the purpose of the evaluation and the composition of the data. In the case of the royalty interest evaluation, the charges must be made against the gas price since there are no other operating costs.

## Deductions from Secondary Product Prices and Income

Secondary products such as NGL, particularly when processing is done by a third-party gas plant or other facility, may also be subject to a variety of deductions and charges. In some instances, a portion of the NGL being processed is taken in kind to cover the cost; the dry gas and the residual portion of the NGL is returned to the producer. Processing fees charged by gas plants for wet gas and for shipping NGL may be deducted from the price. Accounting for NGL processing costs as either deductions from price or as operating costs depends on how the processing costs are treated by the producer. Usually, the only source of data for deductions are statements provided by the processing facility.

# 7

# ACCOUNTING FOR ROYALTY AND OTHER NON-WORKING INTERESTS

Royalty interests and other non-working interests can be a burden on the income stream and the value of a property. As discussed earlier in the section on Real Estate, Mineral Rights, and Ownership Interests, property ownership can be complicated. There is commonly a working interest (WI) owner who makes the investment to develop the property and pays the costs of production. The WI owner receives the large majority of revenue, typically 87.5%, from the property. There may be multiple WI owners, but for simplicity assume there is only one. The WI is usually a leasehold interest; that is, the WI has a lease on the oil & gas rights (Lessee). The Lessee has only the right to produce the oil & gas according to the terms of the lease. Where the WI owns the property in fee, his ownership is 100%.

The mineral rights owner is generally known as the RI owner since he receives a payment or "royalty" from the Lessee in exchange for the leasehold right to produce oil & gas from the property and receive revenue from the sale of that production. The amount of the royalty is usually calculated as a percentage of the sales revenue to the property and is deducted from the revenue stream.

When constructing an income stream for a property, it is essential that the various WI and RI ownership interests in the property be defined and accurately applied. For internal evaluations the company Land or Legal department should have current information. For non-operated properties or for properties in which the company does not own an interest, some research may be required.

Obtaining ownership information should not require a title search by an evaluator. The primary source of ownership data that can be used as confirmation of owner information are the **division orders** issued by oil and/or gas purchasers in most states. Division orders are usually the result of extensive searches of title records and are commonly recorded as legal documents in the county or parish where the property is located. However, some states do not require division orders, or the ones on record may not be current. Documentation provided by the owner of the property interest and/or third-party title opinion can be used as a supplemental source of ownership information.

# Royalty Interests

There are several different types of royalty interests that may apply to a property. These include landowner, overriding, and sliding-scale royalties. Regardless of type, royalties are legal contractual obligations that are stipulated either in the lease of the oil & gas rights and which must be paid according to the terms of the lease agreement or in a separate agreement supplementary to the mineral rights lease. The evaluation of a property subject to a royalty interest should not assume that the royalty can be altered or adjusted or that the terms of the lease can be unilaterally interpreted to include deductions from royalty for costs of production not contemplated by the express terms of the lease agreement.

## Landowner Royalty

The Landowner Royalty (LRI) is compensation to the Lessor for allowing access by the Lessee to the resource. LRI can range from the standard 12.5% up to 20% or more. Unless renegotiated, the royalty share is fixed. Accounting for the LRI is relatively simple; the WI is reduced by the royalty percentage interest to a Net Revenue Interest (NRI). If the WI is 100% and the landowner royalty is 12.5%, then the WI is 100% and the NRI is 87.5%. The WI pays 100% of the costs and receives 87.5% of the production and revenue.

## Overrriding Royalty Interest

An Overriding Royalty Interest (ORRI) can be created by the Lessee or other WI owner. An ORRI is often provided to geologists and/or others who play a role identifying the production potential of the property and, in some cases, aid in property development. The ORRI is usually smaller (1-3%) than the landowner royalty and may have a limited duration. An ORRI is separate from and in addition to the Landowner (Lessor) royalty; it is derived from the Lessee's interest and does not impact the Landowner Royalty. The ORRI is a legal obligation that must be considered in an evaluation of the property.

## Sliding-Scale Royalty

Some properties, particularly federal, state, or provincial leases, may be subject to a variable or Sliding-Scale Royalty. This form comes into use when oil and/or gas prices experience frequent changes, or a large-interest owner is concerned about the possibility that a property would be shut-in for economic reasons.

As an example, in the late 1980s, when oil prices declined sharply, the State of California and the City of Long Beach, each of which own a portion of the mineral rights underlying Long Beach city and harbor and retain a royalty interest in a large portion of the THUMS Unit in the Wilmington field, negotiated a sliding-scale royalty system wherein the royalty rate would be adjusted downward from the contract royalty rate if oil prices declined to certain levels. This adjustment had the effect of allowing production, which might otherwise have been shut-in, to continue and to provide revenue to the operator and to the state and city.[16]

Incorporating a sliding-scale royalty into an evaluation can be difficult, but if there is potential for the situation to occur, the sliding-scale royalty must be considered.

## Step-Scale Royalty

Another form of royalty is the Step-Scale Royalty where the royalty rate increases in steps, usually tied to an increase in average production rate. For example, the royalty rate might be 12.5% for the first 20 BOPD of production, 16.667% for the next 30 BOPD, and so on. This form of royalty is rare but could have application to new development or EOR projects where the landowner and/or mineral rights owner perceives increased risk to his property and requires enhanced protection in the form of increasing royalty. The step progression requires some dexterity in constructing the production projection and royalty share calculation but otherwise presents no particular difficulty.

# Non-Working Interests

Another form of non-working interest ownership is a **Net Profits Interest (NPI).** In an NPI agreement, a third-party obtains a share of any profit or positive income stream that occurs from the property. An NPI is similar to a royalty and may come in many forms, but whereas a royalty is paid on revenue before operating costs, the NPI is generally paid as a portion of the net income remaining after *operating costs*, and, in some cases, *after capital costs*. This form of ownership is not common but occurs when a WI owner is unable to contribute his share of capital expenses, or a Lessor has negotiated a particularly good agreement. In the first case, the NPI is similar to a carried WI in that it may have a fixed term based on either production or, more likely, cumulative revenue.

As an example of the second case, the mineral rights owner of a large property in Kern County, who already had a conventional landowner royalty of 20%, negotiated an additional 16.667% NPI for the same property as compensation for allowing unitization of the property for a gas-injection pressure-maintenance project.

The NPI may be cumulative or non-cumulative. The NPI is termed **cumulative** if the capital investment and/or any negative income stream must be recovered out of future profits; during that time the NPI owner receives no income. For example, if the WI owner incurs a usually large expense of $1,000,000 in April and the monthly average income after operating costs is $50,000, it might require 20 months to recover the expense, during which time there would be no "net profit" to be paid. NPI payments would resume when the expense is fully recovered. The NPI is commonly managed and paid by the WI owner. The NPI is termed **non-cumulative** if the recovery of capital investment and/or any negative income stream must be recovered out of future profits in a specified time frame (i.e., the current fiscal or calendar year); during that time the NPI owner receives no income. Net income would resume at the end of the specified period.

Net profits interests can be difficult to evaluate, but must be considered where applicable. The calculation is a two-step process. The first step is to estimate the potential income to the WI after operating costs, taxes, etc. – the net profit (undiscounted). If the NPI is cumulative after investment, then any investment amount must also be deducted from cash flow before calculating net profit. Once the net profits share is determined, it must be deducted from the WI Income before calculating the economic benefit of the WI. The share of the NPI can be calculated as a share of the total net profits. The calculation of the effect of NPI and/or of the NPI interest is made complicated if the income stream is projected to have periods of lesser or greater (or no) income due to anticipated increased costs or investments.

| Year | Net Cash Flow, $ | Investment, $ | WI Net Profit, $ | NPI (12%), $ | WI Net Cash Flow, $ |
|---|---|---|---|---|---|
| 1 | 500,000 | 750,000 | 0 | 0 | 0 |
| 2 | 400,000 | | 150,000 | 18,000 | 132,000 |
| 3 | 300,000 | | 300,000 | 36,000 | 264,000 |
| 4 | 200,000 | | 200,000 | 24,000 | 176,000 |
| 5 | 100,000 | | 100,000 | 12,000 | 88,000 |
| Total | 1,500,000 | 750,000 | 750000 | 90,000 | 660,000 |

**Table 1.** Working interest cash flow projection example

# 8

# COSTS OF PRODUCTION – OPERATING COSTS

An oil and/or gas property has economic benefit and value only so long as it can produce at a profit; that is, so long as revenue from production exceeds the costs of obtaining the production. Once the economic limit of production is reached, the property ceases to have value. One of the most difficult parts of many property evaluations (and often the weakest) is the derivation and application of the costs of production. There are at least three reasons that a thorough operating cost analysis is necessary as part of a property evaluation.

- Estimation of the economic limit of production – Economic limit is a function of production rate and oil/gas price, but it is controlled by the costs of production. Determination of appropriate operating costs for a property is essential. There can be differences in value for a property if the economic limit is determined to be 10 years rather than 20 years from the date of evaluation. An erroneous derivation of operating costs can alter the estimate of economic benefit of an oil property by causing net revenue (gross revenue minus costs of production) to appear to be higher or lower than is appropriate, and/or by causing the economic limit to be extended or reduced, thereby indicating an erroneous estimate of value.
- Analysis of the effects of certain costs on property value – A thorough cost analysis could allow definition of those costs that have particularly large impact on the economic value of a property and provide an opportunity to assess how changes in those costs could affect property value.
- Definition of areas of high costs and/or potential operating problems – Aside from use in the economic evaluation, a comprehensive cost analysis could illuminate areas of high costs that could be reduced for ongoing operating purposes and suggest areas of property operations that could be considered problems requiring correction.

# Data Sources and Collection

The best source(s) of data are the costs incurred on the property being evaluated by the current operator, and, to a lesser extent, the costs incurred on similar properties, assuming, in both instances, that the information is available. Obtaining operating costs from the current operator is necessary because every property is different and will incur expenditures that are specific to both the property and the operating system in use. On the other hand, obtaining cost information from other properties can be useful as a means of measuring the reliability of the subject property cost data and suggesting whether those costs are representative or unique. Occasionally, operating costs data is available through third-party studies.

The time period of data coverage is a function of the operating history of the property and the type of operation. As with product prices, operating costs may be subject to short-term increases or decreases, which may be misleading if not placed in a longer-term context. Where production and operations are relatively stable, a 24- to 36-month history of monthly cost data should be sufficient. However, where operations are subject to change, when development is ongoing, or where the operator has recently changed, a longer period of data coverage (48 to 72 months) may be necessary in order to adequately define both the full range of costs and the variation of those costs with time and production changes. Data provided on a monthly rather than annual basis is preferred so that patterns or cycles in costs can be defined. This level of detail may appear daunting and difficult to obtain, but, in an age of computer-based accounting and spreadsheet programs, absent any other impediments, the task can be manageable.

The level of detail in the cost data must be sufficient to allow a rational projection of future costs consistent with the expected operation of the property. Lump sum costs of so many dollars per year or per month or dollars per barrel produced are generic, do not provide sufficient specificity, and, consequently, have limited utility in an economic evaluation. The costs should be broken down into as many categories as possible so that they can be allocated properly.

When asked for the cost of production for a property, an operator may suggest that the costs are X dollars per well or Y dollars per barrel of oil. If the costs are $1,000 per well per month and there are 30 wells then the total costs are $360,000 per year: a reasonable start. However, would the costs remain the same if the well count dropped from 30 wells to 20 wells or would the costs decline and, if so, by how much? Not all costs are a function of well count. Some costs might decline (contract labor, pump changes) but other costs (salaries and wages) might remain the same or decline, but to a lesser extent.

Similarly, costs defined as dollars per barrel are misleading, particularly if the relation is to barrels of oil produced. Assuming, as above, total costs of $360,000 per year, would the costs change directly in proportion to the oil production? Are pumper wages a function of production or are wages paid as dollars per month regardless of production? Another thought: Assume a property that produces water along with oil and that water production is increasing relative to oil production (WOR). If operating costs are quoted as a function of oil production alone, the total costs would appear to go down even when total production (oil plus water) is increasing. In this case, the lump sum cost is inadequate.

Obtaining cost data is often a test of wills between evaluator and data source, whether the source is a client, the current owner or operator, or a third party. Operators are often reluctant to provide cost data in sufficient detail and/or over a long enough period to allow a useful analysis. Familiarity with the property and the operator is often helpful when combined with a modicum of persistence and diplomacy.

# Analyzing Production Costs

The costs of oil & gas production include all the expenditures required to produce, process, treat, and sell the oil, gas, and any other products from the property being evaluated. There are many different types of costs, and companies may account for the costs in various ways, but to the extent possible, and for the reasons enumerated above, they must be included in the construction of the income stream. It would not be unusual for the operator to agree to provide data but in an accounting format that fits the company requirements for financial reporting and/or corporate tax purposes; however, in that case an effort must be made to understand the entries and, if possible, obtain a breakdown or sub-accounting presentation. If a company provides an accounting format set of cost figures titled "Field Charges," where that line item accounts for 80% of the total costs, the Field Charges must be broken down into the sub-accounts.

Given enough data, operating costs could be subdivided into at least six cost categories that should be considered and derived from the detailed data as part of an evaluation:

- Fixed Costs, $/Month
- Well-Related Costs, $/Well/Month
- Variable or Volume-Related Costs, $/Unit Production
- Periodic Costs, $
- Fuel Costs, $/Unit Purchased
- Overhead, % of Revenue or % Non-Fuel Costs or $/Month

## Fixed Costs

Fixed costs are those which remain the same from month to month. These costs do not change as a function of well count or production volume. They may increase or decline over long periods as operational changes are implemented. However, for income stream purposes they may be considered fixed costs. These costs are scheduled as dollars per month since payrolls and similar costs are accrued on a monthly basis.

- Property/Lease Labor – Labor costs include salaries and wages for all company and contract personnel who work *continuously* on the property and contribute to the operation of the property, such as (some old terms) production foreman, pumper, testing and treating personnel, and field maintenance staff. Even properties that have been automated and may be integrated into a larger operation require someone to monitor activities, define and fix problems that occur, provide tests of the automated system to ensure proper operation, and monitor production and shipping activities.
- Engineering/Supervision – Salaries and wages for engineering and/or supervisory staff that is assigned to the property and maintain a presence on the property, that is, the production engineer at Kern River, not in Houston.
- Costs of Maintenance and Repair of roads, tank batteries, gathering lines/pipelines, general use facilities.
- Vehicles Maintenance – Cost of maintenance for trucks, automobiles, and other equipment, whether done by onsite personnel or by vendor services.
- Rents for office space, office and field equipment, staff or field vehicles, and rents to surface owners and/or third parties for access for roads, pipelines, etc.
- Insurance – Mechanical and liability insurance covering operation, staff/employee insurance where applicable and which is not part of salaries and wages.

These costs are accumulated and converted into an average monthly cost expressed as $/Month. Where a long production life including a declining number of wells and/or production volume is projected, some consideration should be given to periodic changes in fixed costs to reflect reduced activity.

## Well-Related Costs

Well-related costs are usually any costs that are directly incurred to keep individual production and, where applicable, injection wells functioning properly.

- Third-party Contract Labor
- Equipment, Materials, and Parts
- Pump Repairs and Replacement
- Well Pulling and Maintenance
- Repair of Flow Lines and Meters
- Idle well tests

These costs may be accumulated over one or more years and are scheduled as average dollars per producing well per month ($/well-month). Where appropriate, and where data is available, the costs of injection wells could be derived separately as dollars per injection well per month. Incorporating per well costs into the evaluation requires a projection of the number of producing wells and injection wells that are expected to remain in service. Depending on the location and age of the property, the costs to maintain idle wells, such as shut-in pressure tests, may also occur and should be included as a function of production wells costs.

## Variable or Volume-Related Costs

Variable or volume-related costs (sometimes referred to as "*lifting costs*") are directly related to the total volume of fluid produced and vary with the volume of gross fluid (oil plus water) production. The cost to lift a barrel of water is the same as the cost to lift a barrel of oil. The unit cost is calculated as $/Barrel of gross fluid, not $/Barrel of oil, and is applied to the expected gross fluid production. If the property produces water and the WOR can be expected to increase while oil production declines, then applying the costs only to the oil volume will overvalue the property. A reasonable projection of future water production as a function of oil production will be required to complement the oil production projection and to provide an estimate of gross fluid production. Lifting costs can include:

- Electric Power/Natural Gas Fuel – The cost of power to run pumping units or gas-lift systems is usually one of the highest cost categories after salaries and wages.
- Emulsion Treating and Separation – The costs of heating oil/water separation vessels and, where necessary, emulsion breaking chemicals.
- Wastewater Treatment and Disposal – The costs of chemical treatment of produced water to allow either functional injection or disposal plus the costs of power for pumping equipment to transfer wastewater from tank batteries to the disposal site.
- Corrosion Control – The costs to protect equipment, tanks, flowlines, and other facilities from corrosion.
- Treatment for Sale – The costs of on-site treatment of crude oil and natural gas for sale and the maintenance of shipping facilities.

For properties that are subject to enhanced recovery through water, steam, and/or gas injection, some operators and evaluators subdivide expenses into production volume and injection volume-related costs. This can be a useful approach, particularly for the operator, but can result in additional evaluation work including projection of an injection schedule. This is a rational approach and, where data is available, is a correct method of evaluation. An EOR project run by a competent operator should already have a plan for ongoing injection rates/volumes that can be used in an evaluation.

There is an alternative method. The purpose of EOR injection is to increase or maintain oil production. Under that premise, all costs for both production and injection could be combined and related to gross fluid volumes produced, without the necessity of a separate injection volume cost. The caveat is that such an approach may not be applicable for the entire project, particularly if the project is segmented such as for multiple reservoirs and/or fault blocks. This is entirely a judgement call subject to data availability. Where the injection cost data is available and where applications of unique production and injection costs are feasible, then separate accounting for these costs is valid.

## Wastewater Disposal

Wastewater disposal can be a major variable cost but, in some contexts, deserves to be derived and applied separately. Dumping produced water into sumps or waterways is no longer acceptable. Disposal wells on the property must be developed and maintained. Where disposal on the property is not viable, an approved off-site disposal well or other site is usually mandated, which may require trucking or pipeline shipping at some cost as well as a disposal fee. In some urban areas, properly treated processed water can still go into the sewers, but there can be substantial fees, which are often volume-based, and are considered variable costs. Sewer disposal must be viewed as interruptible and therefore having a limited duration.

- Regulatory Cost – Inspection and testing of on-site wastewater disposal facilities including chemical analysis, treatment of water, and testing of on-site disposal wells.
- Transportation – Where on-site disposal is not an option, there will be costs for trucking or pipeline transportation of wastewater to an available disposal site.
- Disposal Wells – Costs of maintaining and operating on-site disposal wells
- Fees and Unit Costs – Mandated and approved wastewater disposal sites, whether wells, settling ponds, sewers, or other location, usually require payment of a monthly or annual fee to use the site plus a volume-related charge ($/BW) and a fee for periodic water content/contamination testing.
- Limited Duration – Estimated wastewater disposal costs can also require an estimate of the time that the disposal site/system will be available as well as an analysis of alternate disposal options.
  It is not unreasonable to view the ability to manage and/or dispose of wastewater as a criterion for determining the remaining production life.

## Natural Gas Costs

Assuming that there is a measurable and (presumably) saleable volume of gas production associated with crude production, there may also be distinguishable variable/volume-related costs associated with the production, processing, and disposition of natural gas and other products. One criterion might be that, if a product generates a revenue stream, it should also have a cost stream. To the extent allowed by data availability, a separate cost category for produced natural gas should be considered and applied to the gas production. This will require a projection of expected natural gas production and income.

- Gas Separation from Crude – Depending on the volume of gas relative to oil and the size of the required facilities, a charge for gas separation may be applicable.
- Dehydration – The costs associated with removal of water from produced gas.
- Separation of Liquids – If separation takes place on the property in a gas plant or simple chiller unit, there could be measurable costs depending on the volume of gas and size of the facility.
- Shipping – Costs for maintenance of any on-site gas shipping facilities – meters, pumps, and gathering lines.
- Compression – Costs associated with compressing gathered gas for transportation to a sales pipeline and/or reducing downhole pressure to allow increased production.
- Transportation – This cost is often hard to categorize and pin down as it is commonly a charge from a gathering system operator.

## Periodic Operating Costs

Periodic costs are generally associated with well maintenance and are often sufficiently large as to stand out in operating cost statements. They include activities that do not occur continuously but recur with sufficient frequency that both the timing and amount of the cost are reasonably predictable. Such costs might be:

- Deep well pump changes requiring several days of contract well work.
- Liner replacement in wells that are subject to damage from corrosion/erosion from produced fluids and/or entrained materials.
- Wellbore hot-oiling as might occur in low-gravity asphaltic-base oil wells.
- De-waxing of liners and wellbores such as might occur in paraffin-base oil wells.
- Acidizing, cyclic steaming, or other stimulation or maintenance requirements.
- Periodic regulatory testing of idle production, injection, or wastewater disposal wells.

Periodic costs may be of sufficient magnitude that specific budget authorization is required. In that case, an evaluator must decide whether the costs should be treated as operating costs or as capital investment. The inclusion of these costs should be reviewed as the property nears economic limit to determine whether it is appropriate to adjust the scheduling of the expenditures.

## Fuel Costs

Fuel Costs can and should be a separate category when the source of the fuel and the cost justifies special consideration. This category usually refers to the cost of natural gas to fuel steam generators or other special use facilities and/or electric power for pumping units and other field uses. The governing criteria is that if natural gas and/or electric power are obtained from a third party, whether a utility or another operator, in sufficient volume to impact the property if that fuel/ power were not available. The source of this data should be purchase contracts, and purchase and payment records.

Certain special cases regarding fuel source require mention. If produced oil and/or gas is retained for use on the property as lease fuel, commonly referenced as Used-on-Lease (UOL), the volume used for fuel can be included in the income stream by reducing the amount of production available for sale, or by deducting the equivalent value of the fuel volume as an operating cost. The UOL production serves an economic purpose even though it does not produce revenue by substituting for costs for fuel that would otherwise have to be purchased. The choice of treatment usually depends on whether or not the lease agreement requires payment of royalty on production used as lease fuel (most leases do not) or whether fuel usage can be deducted for production tax calculation (generally, no). Deduction of fuel usage from production may distort the income stream; therefore, the best approach may to treat produced fuel usage as a separate operating cost subject to the prevailing price for oil or gas used. Another judgement call.

## Overhead

Overhead is considered carefully, whether valuing a single property or multi-property project. Overhead is often an indirect expense that is incurred off the lease but is then allocated to the property. Such costs might be district or head office supervisory and engineering staff. Overhead also includes distributed costs of environmental, regulatory, or other programs that are applied across several properties or which have no property-specific application. The overhead to be considered should only be the cost of direct management supervision of the property and staff services related to the property. The usual approach is to develop overhead costs as a percentage of recurring operating costs. An alternative is to assign a dollar amount per year. Reference to COPAS data may be helpful.[17] The decision whether and/or how to include distributed Overhead may also be a function of the purpose of the evaluation. [**Note**: Not all properties have need of or incur overhead charges.]

## Ancillary Operations

These include the costs of operating facilities that are necessary to maintain production but which have some of the characteristics of stand-alone improvements. Steam generation facilities, co-generation facilities, gas processing plants, and wastewater recycling and/or disposal systems fit this category. Such systems rarely serve only one property unless that property is particularly large. If the ancillary operations are not part of the property being evaluated, the evaluator must determine if the facility represents a revenue and/or cost component of the property value and allocate cost/revenue appropriately.

# Projecting Future Costs

As noted earlier, operating costs, though not as volatile as oil & gas prices, can be subject to short-term increases or decreases, which may be misleading if not placed in a longer-term context. Operating costs are subject to inflationary and other pressures that may result in changes in future costs. A projection that assumes normal inflation of labor and other material costs will generally serve the purpose of most valuations. Fuel costs may be subject to the same potential changes as are considered for natural gas prices but are also subject to regulatory controls and contract terms. In evaluations where fuel costs are large, research and prudent judgement are necessary before projecting those costs. On the other hand, if there is sufficient data, it may be possible to determine if operating costs for a property are changing by any measurable amount. This approach might be possible for large properties or units with long, accessible cost records but is likely not justified for a small property. When in doubt, use the industrial/commercial core rate of inflation reported by government agencies and/or industry data sources such as the American Petroleum Institute (API).[18]

# Incorporating Cost Data in the Evaluation

Assuming that sufficient operating cost data has been collected and analyzed to allow segregation into some or all of the categories discussed above, and further assuming that the calculation method available, whether pad and pencil or computer software, has sufficient flexibility to accommodate inclusion of the cost data, a relatively detailed evaluation can be built.

- Fixed costs are calculated as $/Month times the number of months (12 or less) times inflation or other annual adjustment.
- Well-Related Costs for production wells are calculated as $/Well/Month times inflation or other annual adjustment multiplied by the number of projected producing wells.
- Well-Related Costs for injection wells are calculated as $/Well/Month times inflation or other annual adjustment multiplied by the number of projected injection wells.
- Variable Production Costs are calculated as $/Bbl of gross fluid times inflation or other annual adjustment multiplied by the volume of projected oil and water production. This application requires a projection of future oil production and a separate projection of expected produced water, added together as gross fluid.
- Variable Injection Costs, where appropriate, are calculated as $/Bbl of injected water (steam injection is water injection with higher costs) times inflation or other annual adjustment multiplied by the projected volume of injected fluid. A reasonable argument can be made that since the objective of injection of water or steam was to increase oil production, the costs of injection should be added to the oil production costs.
- Wastewater Disposal Costs, where appropriate as a separate category, are calculated as $/Bbl of water times inflation or other annual adjustment multiplied by the projected volume of disposal water.
- Natural Gas Production Costs are calculated where applicable as $/MCF times inflation or other annual adjustment multiplied by the projected volume of gas production.
- Fuel Costs, where appropriate as a separate category, are calculated as $/MCF for gas and/or $/MWH for electric power, adjusted as necessary by contract terms or other source, multiplied by the projected volume of natural gas and/or electric power purchases.
- Overhead is commonly calculated as a percentage of fixed costs or combined costs. Some discretion is required for overhead, depending on the property and the purpose of the evaluation.

# 9

# SEVERANCE, PRODUCTION, AND PROPERTY-RELATED TAXES

Production and property-related taxes, fees, and assessments are commonly imposed by state governments but can also be levied by a county, city, or other jurisdiction within a state. They can and do vary in form and function and in how the taxes are described and calculated. These impositions include, but are not limited to, severance taxes, production taxes, wellhead taxes, regulatory fees, and ad valorem taxes. These taxes and levies occur individually but, in many areas, one or more of the taxes may occur simultaneously and are cumulative. Certain types of taxes may be imposed and managed by different agencies. Taxes are created and applied by government entities to raise funds for government purposes and have been known to change over time in both amount and complexity, and generally to increase. It is prudent to identify the taxes applicable to the property being evaluated and to include them in the income stream.

**Note:** Tax rates discussed in this section are taken from information compiled from different sources in studies from 2011, 2014, and 2018, as well as some current data obtained from various agency websites. The information is subject to change in both the form and the specific tax rates. The following examples are taken from only the more prominent producing states. Further, the information is not exhaustive and is offered only as illustration to encourage recognition of the existence and effect of taxes on economic evaluation. Reference should be made to the appropriate agency in the state and/or local jurisdiction.

## Severance Tax

Several major producing states impose a severance tax on oil & gas production. The tax is calculated as either a percentage of the sales income before any costs, including royalty or other taxes, are deducted, or a levy of so many dollars per barrel of oil or MCF of gas (sometimes referred to as a Gross Production Tax). There are some exceptions. For this discussion only the percentage of sales revenue is classed as severance tax. The tax may be a different percentage for oil and for gas. Severance taxes are commonly paid by the product purchaser and reported back to the product seller as a reduction in sales revenue.

**California** – As of 2023, California does not have a severance tax; however, the state does allow cities and counties to impose "severance" taxes as they determine appropriate. These local taxes are more in the nature of Gross Receipts taxes or Production taxes and are discussed below as Production taxes.

**Louisiana** (2021) – For oil severance tax purposes, Louisiana classifies wells as Capable, Incapable, and Stripper and assigns different tax rates to each class. An Incapable well is less than or equal to 25 BOPD and equal to or greater than 50% water cut. A Stripper well is less than or equal to 10 BOPD. A Capable well is one that is not Incapable or Stripper.

Capable well tax – 12.5% of value

Incapable well tax – 6.25% of value

Stripper – 3.125% of value

Value is defined as the higher of gross receipts less trucking, barging and/or pipeline fees, or posted field price. The natural gas severance tax is more in the nature of a production tax. The tax for a capable well is $0.111 per MCF, with many exceptions and variations.

**New Mexico** (2021) – The severance tax on produced Oil and Natural Gas is 3.75%.

There are lower rates for certain classes of production.

**North Dakota** (2021) – The tax is referred to as a Gross Production Tax. For oil, the tax is 5% of gross value at the wellhead.

**Pennsylvania** – As of 2021, Pennsylvania has not introduced a percentage severance tax.

**Texas** (2021) – The oil severance tax is 4.6% of the market (sales) value of oil produced except that the rate is 2.3% for oil from EOR projects. The gas severance tax is 7.5% of the market (sales) value of natural gas produced.

**West Virginia** (2021) – The severance tax on produced Oil and Natural Gas is 10%.

**Oklahoma** and **Utah** – have sliding-scale Severance (Utah) or Gross Production (OK) taxes that change based on the price paid for the oil and/or natural gas. As with sliding-scale royalties, these changeable rates were intended to reduce the tax burden when prices decline but would increase the applicable taxes when prices increased.

The **Utah** oil tax rate (2021) is 3% of value for the first $13 per barrel of oil but increases to 5% if the value is $13.01 or higher. The gas severance tax rate is 3% of the value up to and including the first $1.50 per MCF and 5% of the value from $1.51 and above per MCF. There are numerous exemptions.

The **Oklahoma** oil severance tax is similar to Utah.

Because severance taxes come "off the top," they are usually calculated in an evaluation as a direct deduction against the oil and/or gas sales revenue.

That is, if the severance tax is 4.6%, the tax calculation is 4.6% times gross sales revenue and remaining gross revenue to the producer is 95.4% of the sales amount.

## Privilege Taxes

Some states (Alabama, Mississippi) refer to fixed-percentage taxes as privilege taxes. The functional effect is the same as a severance or gross production tax. For example:

**Mississippi** – For Oil, the privilege tax is 6% on the value at the point of production. The tax is reduced to 3% when oil is produced by an EOR project using transported CO2. For natural gas the privilege tax is 6% on the value at the point of production.

## Production Taxes

Production taxes take many forms depending on the location and jurisdiction. In certain cases, they are similar to and calculated in the same way as severance taxes. The distinction is that they are usually stated as a dollar amount per barrel or MCF. Another distinction could be based on whether the tax is paid by the oil or gas buyer or by the producer. Where these taxes are subject to change, and particularly when the tax is paid by the producer, it may be preferable to treat them as part of the operating costs and use the oil and/or gas price before tax deductions. In many cases, the amounts of these taxes are small and have only minor impact on value; however, where such a tax is known to occur, it should be included in the evaluation. Wellhead taxes are generally a fixed amount such as dollars per well per year.

**California** – Numerous cities, particularly in Los Angeles and Orange Counties, have taxes that are codified as Severance taxes but are Production taxes charged as $/Bbl of oil or $/MCF of gas.

Beverly Hills (as of 1999) – $2,785.12 for 10,000 Bbl or less per well plus $0.27 per added barrel

Culver City – $18 per $1000 gross receipts

Inglewood – $0.22 per barrel

Long Beach – $0.15 per barrel

"Business license fees" or "Fire department inspection fees"

Long Beach – $210 per well per year

Los Angeles – $511 per well per year

Inglewood – $330 per well per year

**Illinois** – (statewide annual "permit fees")
$150 for the first permit
$300 each for 2-5 wells
$750 each for 6-25 wells
$1,500 each for 26-100 wells

**Ohio** – State wide rate of $0.10 per barrel and $0.025 per MCF of gas

## Regulatory Taxes

Regulatory taxes and/or fees can appear similar to severance or wellhead taxes but are usually directed toward specific uses, such as funding a particular state agency or an environmental fund. Such taxes vary greatly from one jurisdiction to another, can consist of either percentage or fixed amount per unit, and they may only be enacted to apply for fixed time periods. Several states and some local jurisdictions have begun to impose taxes or fees to fund "Environmental Remediation" or "Oilfield Site Restoration." These taxes and fees may be variable over time, particularly when the intent is to accumulate a certain amount of funds.

**California** – has a legislated tax of $0.5653369 (2019-2020) per Bbl of oil production or per 10 MCF of gas production, imposed by the State of California to support the Division of Oil, Gas, and Geothermal Resources (now Geologic Energy Management – GEM).

**Texas** – has an Oilfield Cleanup Regulatory Fee of $0.00625 per barrel and $0.0007 per MCF until the fund reaches or exceeds $10 million.

## Ad Valorem Taxes

Ad valorem taxes are property taxes, similar to the assessed value tax on a home, farm, or business location. In theory, the taxes are based on the (Fair) Market Value of the oil & gas income attributed to the property. The value of specific properties is determined by the taxing authority such as a county tax assessor (California) or appraisal district (Texas) using an annual or bi-annual economic valuation of future production and income. In both states, each property is revalued as of January 1 of each year.

Where ad valorem taxes are imposed, they can be an important cost element for producers. In California, where there is no severance tax, the ad valorem tax can equate to as much as 3-5% of gross revenue. In some jurisdictions, such as Kern County, the oil & gas property tax is an important source of county revenue.

**California** – the State does not conduct the assessments; however, valuation methods to be used by the counties are promulgated by the State Board of Equalization ("SBE") and are published in several manuals of instructions, including *Assessors Handbook 566*, which specifically lays out methods and procedures for appraisal of oil & gas properties by County Assessors. The SBE instructions are commonly written with the participation of county assessors and industry experts and are supported by the outcomes of previous litigation of petroleum property assessment and taxation. The litigation and subsequent legislation have resulted in several SBE Rules, such as Rule *8* and Rule *468*, that define the extent of oil & gas property appraisal for ad valorem tax and the methods to be used. Property tax rates in California, for all properties, are governed by Proposition 13 (an amendment to the California Constitution passed in 1977), which set the tax rate at 1% of assessed valuation; the assessed valuation is limited to a 2% annual increase in the base value. If the property is sold, the purchase price establishes a new base value.

**Texas** – takes a less formal and documented approach but does maintain a Property Tax Division of the Comptroller's Office to review and comment on ad valorem tax assessment and valuation in Texas. Texas has published a *Manual for Discounting Oil and Gas Income* for use by Appraisal Districts. Further, Texas publishes an annual *Oil & Natural Gas Price Escalation Forecast* that is supposed to be used by Appraisal Districts. The tax rate varies among the many appraisal districts.

The difficulty with ad valorem taxes is that the assessed property values do not always follow changes in property value caused by variations in economic conditions. In both states, large production companies have dedicated property tax departments to manage tax issues; smaller companies employ tax agents or consultants to negotiate tax questions and values with county assessors or appraisal districts. Both states have local industry groups dedicated to the review, discussion, and negotiation of tax issues. The ad valorem tax is relatively minor in most other states and, where it occurs, applies primarily to surface equipment (Louisiana) and, in some cases, the value of the wellbore. It may be necessary to review assessment records and/or prior tax statements to estimate future property taxes. Ad valorem taxes can be converted to a percentage of gross revenue or deducted as an annual amount.

## Sources of Tax Data

Production, severance, regulatory and other taxes and fees change with time and from one jurisdiction to another. The trend in government at all levels is to increase the amount and complexity of taxes and fees. The examples cited above are from surveys taken at different times before this text was written. The best source of current and applicable tax information is the operator of the property and/or the purchaser of the crude and/or natural gas. A secondary source is the state and locality where the production is located.

# 10

# ECONOMIC LIMIT – CONCEPT AND APPLICATION

For valuation purposes, the estimation of potential production and income requires that there be an end point. The end point can be an imposed limit, such as the flow rate at a certain wellhead pressure for gas wells, or an oil production rate, or a period of time, such as 15 years. However, for valuation purposes, the end point is most commonly a form of economic limit of production.

## Concept

The economic limit of production, also known as the economic **break-even point**, is the point in the production life of the property at which the income received from the produced oil, gas, and other products equals the cost of producing that income. That is, if the cost of producing and selling 10 barrels of oil is $100, and the income received from the sale of the 10 barrels is $100, the property has reached the economic limit. In common usage, economic limit is often expressed as a production rate. The period of time from the effective date of the evaluation to the economic limit is the remaining **economic life**. Petroleum properties deplete and production generally declines so that a projection of production to an economic limit is not particularly difficult. While simple in concept, the economic limit is often difficult to define in an evaluation.

Historically, the Economic Limit (EL) was estimated by a rather simple and generally accepted relation, where:

$$\text{Economic Limit} = \frac{\text{Total Operating Cost}(\$/\text{Month})}{\left[\left(\left(\text{Oil Price}(\$/\text{Bbl})\right) - \left(\text{Production Related Taxes}(\$/\text{Bbl})\right) \times \left(1 - \text{Royalty}, \%\right)\right)\right]}$$

$$= \text{Bbl/Month}$$

The economic limit is the point where the total costs of production divided by the working interest (WI) share of production income equals a certain production rate.

However, the relation illustrates some of the complexity of the concept and problems of application. There are several components to be considered in this seemingly simple equation. The primary difficulty is with the numerator, Total

Operating Cost (TOC). The equation assumes a Total Operating Cost expressed as $/Month. But as discussed in the section on operating Costs, there are several categories of operating costs that are functions of production or well count; not all costs are fixed or tied to the calendar. On a small property, where the fixed costs constitute the largest part of total expenses, the cautious use of a single TOC amount might be possible, but such situations rarely occur. The problem expands as the property operation and, therefore, the costs of production become more complex; the simple equation loses traction.

As discussed previously, operating costs vary in composition as fixed, per well, variable, etc. Combining the fixed, well-related, and variable costs into a single amount ($ per month) must consider the relationship of the costs in each category. Assuming that all the costs could be reduced to a single amount that is representative of the actual costs of operation, regardless of the size of the production but particularly for a complex project, is simplistic and defeats the purpose of economic evaluation. While it may be the case that, as noted above, fixed costs are or can be the largest cost category, the variable costs usually run a close second. Treating variable costs as fixed ignores the fact that changes in costs do occur as production changes. In a similar manner, well-related costs may change as the number of wells increases or, in times of declining production, decreases. A single cost element expressed as $/month cannot account for the function of variable or well-Related costs.

The denominator of the equation is less of a problem but must still be considered carefully. The equation uses Oil Price, but revenue may be derived from gas sales as well so maybe Total Product Price, expressed as $/BOE would be better notation. Also, the price should be net of any deductions from the sales income, such as transportation, treating, or other sales-related costs, resulting in a Net Total Product Price, $/BOE. Production-Related Taxes should be read as any tax that comes directly out of the sales revenue, particularly if paid by a third party, such as the buyer of the oil and/or gas. Removing the royalty share of income confirms that the economic limit being estimated is that of the WI since the royalty interest does not pay operating costs.

One option to introduce some flexibility might be to reduce the Net Total Product Price ($/BOE) component in the denominator by the amount of the Variable Costs ($/BOE). The denominator then becomes:

[Oil Price ($/BOE) – Variable Costs ($/BOE) – Production-Related Taxes ($/BOE)] * WI%

This still requires converting all the other costs to a $/Month basis. Further, the equation is static in that it does not allow for future changes in operating costs, production taxes, and/or oil & gas prices over time.

The economic limit concept and the structure of the equation attempt to estimate the production rate at a time when costs of production (in the broad sense) exceed the revenue from production; however, it does so using the costs and prices that exist today. The concept is broad enough to include the idea of an economic limit derived from future costs and product prices, but the equation is not so flexible. Presumably, an estimate of the costs of production at some point in the future could be calculated; in a similar manner, a future product price could be constructed, but doing so requires an assumption of a point in the future which would, by definition, be the economic limit.

## Practice

The concept of economic limit and the reality of reaching an economic limit often differ; when possible/practical, changing conditions should be considered. As a property nears the end of economic production, operations can be revised and operating costs can be adjusted (reduced), in order to extend the productive life. Oil price is variable; an increase or decrease in price can extend or shorten the economic life. Sliding-scale royalty rates can sometimes be negotiated to stave off the inevitable. These considerations are all part of the evaluation process and can be recognized and appropriately included in the valuation. However, the extension or shortening of the economic life of 10-15 years by a few months or years may have little impact on the economic benefit of the property.

In reality, while the concept of an economic limit is fundamental to evaluation, the calculation of economic limit using a simple static relation is outdated. Evaluation methods which allow calculation of literally hundreds of income streams using variable production, price, and cost data can be used to define the economic limit under all the relevant conditions as the point where the stream of net income after all costs and deductions becomes negative. That is, regardless of the size and complexity of the project/property, a spreadsheet can be constructed to include all sources of production and income, and subject to all categories of operating costs exhibiting changes over time until the amount in the monthly or annual net income column becomes negative and thereby reaches economic limit.

As a practical matter, operators are generally reluctant to shut-in production simply based on the economic conditions prevailing at a certain time. A property may be kept in operation beyond the apparent economic limit for a variety of reasons including but not limited to:

- Anticipating improved production and/or operating conditions.
  Wells and/or properties that appear to be uneconomic under existing conditions may be kept either in production or temporarily shut-in while methods of improving production are considered or application of advances in completion/production methods are evaluated.
- Anticipating increases in product pricing or waiting out price declines.
  Oil & gas prices can be volatile. Properties that are operating near the

economic limit may be more susceptible to price declines than other properties.

As an example, following the major oil price decline in 1986-87, many wells on properties in California and elsewhere were shut-in with only one or two wells remaining in production until oil prices finally increased.

- Anticipating reductions in operating costs such as by consolidating and/or improving surface facilities, reducing fixed costs, improving variable costs using different fuel sources, disposal methods, etc.
- Integration of the property into a larger operation, such as combining multiple leases or unitization. Property integration has been an important factor in retaining many properties in production with the objective of improving efficiency, reducing costs, and consolidating disparate operations.
- It is not unusual for an operator to continue production of a property (or wells) even though uneconomic if the property serves a useful purpose, particularly when integrated with other properties. The oil & gas produced may be uneconomic for the property but adds to the total volume produced for the integrated properties.
- Retention of a lease, access to property. Shutting-in the producing wells on a property could, depending on the terms of the lease, trigger a termination of the leasehold and loss of access.
- Regulatory avoidance (Abandonment of Wells). In many jurisdictions, California being one, regulations require that wells that are shut-in (idle) for a period of time (i.e., 5 years) must be abandoned. Even though a property may be uneconomic, the loss of income for a short period of operation can be, and likely is, less than the cost of abandoning wells and facilities.

Whether an economic evaluation of a property should incorporate these possibilities depends in large part on the purpose of the evaluation. Internal evaluations that may include "what if" scenarios could be constructed to account for any or all the above conditions.

The fact remains, however, that at some point petroleum properties become uneconomic to operate and confer no economic benefit on the owners. Recognition of this point is one of the functions of economic evaluation.

# 11

# CAPITAL EXPENDITURES

Evaluation of an oil or gas property must define and consider anticipated capital expenditures if those expenditures will be necessary to achieve the projected production, maintain existing production and/or efficient operations, or to comply with reasonably foreseeable regulatory requirements. If additional production and future recovery are included in the evaluation derived from new drilling, remedial work, or enhanced recovery, there must be sufficient capital investment scheduled at the appropriate time to bring about the changes in production. If drilling of new wells or re-drill of existing wells is planned to increase or maintain production, the required investment must be included in the evaluation before the production becomes effective. Where installation of new pumping equipment is needed to handle increasing gross fluid production, or if compression is necessary to reduce operating costs or to make more gas available for sale, then these expenditures must be scheduled in the evaluation at the appropriate time.

The difference between periodically recurring costs of operation and required capital expenditures is not always clear. Each company may have its own definition. Capital expenditures can be broadly defined as infusions of capital, which may or may not be predictable, and which are generally expected to result in an increase in production or a reduction in costs or both. Changing a down-hole pump might be an operating cost, while replacing a pumping unit could be a capital expenditure. Cyclic steaming might be a periodic cost, but hydraulic fracturing might be an investment. There is a certain amount of subjectivity involved in the categorization of these costs.

Any anticipated capital investment should be included in the income stream and evaluated for appropriate return. Capital expenditures are deducted from net operating income after all royalties, operating costs, and production taxes.

## Sources of Data

The source of investment information generally depends upon the purpose of the evaluation. Evaluations done for internal purposes, such as capital budgeting are, in large part, a function of planned investment so the amounts should be readily accessible. Similarly, evaluations done for working interest WI owners in a property should have access to planned investment information. Where an evaluation is being done for a non-working interest or for regulatory purposes, investment data could be obtained from the operator, when possible, or from third-party sources such as consulting services, industry data bases, and/or government agency databases. Multiple sources should be used where practical.

## Scheduling of Investment(s)

An income stream built over a period of time and including changes in production, operating costs, or other parameters occurring as the result of investment should schedule the requisite investment prior to the effective event with allowance for the purpose of the investment to take effect. Many software programs as well as spreadsheets can be built to enable monthly data entry. In the case of scheduling investment for a new well that takes three months to drill, the capital expenditure must occur at least three months prior to the start of production.

## Initial Investment

Define and tabulate costs for new or replacement wells and equipment, surface facilities, and any other capital expenditures as of the date of evaluation. Those investments which will occur within a relatively short period after the effective date should be accounted for at the current cost.

## Escalation of Capital Investment

Capital costs should be considered subject to inflation and other pressures for change. While general estimates of inflation may be useful, and just as likely to be correct as any other estimate, the potential for changes in the cost for major equipment purchases and contract drilling should be reviewed, when possible, with representative suppliers and vendors.

# 12

# REGULATORY COMPLIANCE AND ENVIRONMENTAL LIABILITY

Regulation of property rights is, by definition, a restriction on the use of property. Zoning controls the potential use and form of development of surface and, in some cases, sub-surface rights. Water Quality Districts, Air Resources Boards, and similar organizations can impose conditions in the form of licenses and permits defining and approving, thereby limiting, activities on a property. Regulatory controls on oil & gas development and production operations can be imposed by local, county, state, and Federal legislatures, as well as by agencies designated by those bodies. While regulations may be promulgated by a wide range of political entities, regulatory compliance costs, regardless of source, are local, apply to specific properties, vary geographically, and have been known to change as the political environment changes.[19]

## Regulatory Compliance

This is an area where serious consideration must be given to future costs. Regulation that impacts property rights, directly or indirectly, positively or adversely, regardless of source, has an effect on property value. An evaluator of a specific property must satisfy two conditions regarding regulatory compliance.

1. Is the subject property operating in compliance with existing and foreseeable future regulations?
2. What are the costs of maintaining compliance with existing and potential regulations? A knowledge of the compliance component of operating costs may help in understanding the subject property.

The immediate effects of regulation on value are measured by the difference in income and value before and after implementation of the regulation. Immediate effects can include the repair and/or replacement of offending equipment, the cost of training and equipping of field personnel to carry out imposed procedures, the rebuilding of treating facilities and storage tanks to comply with air emissions rules, or the lining and/or covering of sumps. These can be major investments that do not necessarily improve production and operations but which come about through regulatory actions. They can have an immediate impact on operations because the order to make changes may carry a short time period for compliance. In many jurisdictions, non-compliance can result in shut-down of the property, thus terminating production and income.

## Long-term Regulatory Costs

These costs are encountered in petroleum property evaluation because they impact the income stream as continuing charges against income, and/or reduction in the property value due to delays in construction, project startup, or full-scale operation. Charges against income result from increased personnel salaries and/or contract services, air and/or water emissions test services, the cost of wastewater disposal, fees for filing for approvals, disposal fees, inspection fees, licenses, permits, and, sometimes, payments for air and/or water quality offsets.

Regulatory issues reduce the value of oil properties when development is impeded by compliance requirements. If a reliable estimate for the completion of the permitting process for a new gas plant is 2-3 years, that time has to be built into the appraisal. If a new drilling project to access a hitherto un-producible oil accumulation requires not just an Environmental Impact Report (EIR) but public approval, then the project cannot be valued as if it were starting up today. Depending upon local attitude, some degree of uncertainty may be prudent. In valuing these properties, there must be allowance for both time delay and the uncertainty of a positive outcome.

> As an example, evaluation of the development potential of a steam injection project in an old, heavy-oil field in Santa Barbara County, California requires that a 4- to 6-year delay in project start be factored in for the local permitting process, coupled with California Environmental Quality Act (CEQA) environmental quality review.

It is not uncommon for regulatory expenses to be overlooked in estimating economic value because the costs of regulation are difficult to estimate and the scheduling of such costs is often subject to third-party (regulator) discretion. While some compliance costs become absorbed in operating expenses and overhead, the reduction in value of a property or project as the result of the delay in initiating construction and beginning production due to lengthy review and approval processes is often difficult to predict. A realistic evaluation, particularly of large properties and projects in the process of development, must consider and include the appropriate costs of both ongoing and anticipated regulatory compliance.

The extent to which the evaluation of a property should include estimates of compliance costs is discretionary to the evaluator and depends upon both the scale of property/project being evaluated and the purpose of the evaluation.

Evaluation of a multi-property steam injection project or expansion of a Marcellus or Eagle Ford drilling program for internal budgeting purposes should require inclusion of all foreseeable and otherwise likely regulatory costs. Aside from the immediate expenses and investments, regulatory compliance costs usually impact the income stream through four broad cost categories.

## Disposal Costs

These costs were discussed above as part of operating expenses where the focus was the ongoing cost of injecting or otherwise disposing of water that is produced along with oil & gas. Some additional disposal costs may be the result of regulatory requirements such as specific chemical tests and treatment costs, or pressure tests for disposal wells. Further, in the context of regulatory compliance and the next issue, environmental liability, disposal costs can also include removal and disposal of contaminated soil or contents of settling ponds and sumps, along with replacement of contaminated equipment. In some jurisdictions all contaminants (variously defined) must be disposed of as they are encountered.

## Labor Costs

Labor costs related to regulatory compliance do not vary with production or well count. It takes just as long to prepare and present reporting to the local Air Quality Management District (AQMD) when production is declining as it does when production is increasing. The Water Resources Board will not reduce oversight simply because oil prices are down. Compliance procedures must continue to be followed, or penalties can and will be imposed. Depending on the size of the property or the complexity of the project, compliance staff will have to be assigned to these tasks. These costs should go into the evaluation as an element of Fixed Operating Costs.

Regulatory compliance costs are also an increasing part of overhead, especially for properties involved in EOR and/or large-scale development projects. The salary for the staff environment specialist and the safety engineer assigned to the project – or, for larger projects, the public relations and other staff needed to interact with both regulatory agencies and the public – are overhead. These can, and usually do, include the time and effort required for meetings and hearings with various "interested parties" such as environmental organizations or homeowners' associations, education of local press, and participation in local, influential service organizations.

Overhead could also include contracting of the archaeologist that must be hired to survey the site, followed by the costs of establishing green-belt space and planting screening trees, landscapers to maintain the green-belt and trees masking the property, along with site redesign to accommodate visual and other impacts;

new roads, schools and other facilities, and flora and fauna inventories must be included in the project budget in order to achieve regulatory approval.

Finally, there is capital investment related to regulatory compliance. There must be a determination of the amount of investment required for a project, in excess of the amount indicated by engineering design criteria, in order to comply with existing and anticipated regulations. This would include everything from the investment for new equipment required to comply with the regulations to the extensive Environmental Impact Reports (EIR) that must be completed and filed with various state and local agencies. It is not unusual for agencies considering approval of a project to impose investments in facilities that are not part of the project as a function of obtaining approval.

> As an example, Santa Barbara County required the construction of a new school for local students as a condition for approval of the Pt. Arguello onshore production facility.

With enough data, and drawing on local experience, an evaluator can estimate the costs associated with compliance, apply them against the income stream, and estimate the future production and income of a property based on the prevailing and expected regulatory environment. If a property has been operated in compliance, the costs should already be identified as part of the operating cost date and no additions may be necessary. On the other hand, if the evaluator determines that the property is not operating in full compliance, or if new development is anticipated, additional cost components may be needed. While data regarding such costs may be available, contracting of a local expert may be necessary.

## Political Risk

In some regions of the country, the difficulty of obtaining approval for a low-impact drilling project, let alone a large-scale EOR project, is such that it is not uncommon to consider that political risk applies. This is a subject that used to only relate to investments in foreign countries but, in recent years, political activities, largely inspired by environmental issues but expanded into an anti-petroleum industry context, now require that the risk of attempting a project in a particular county or state be considered relative to the same investment in another region. These activities have generally taken the form of slow-walking of permitting and approval processes extending to years and/or restrictions or outright bans on drilling in certain areas. Examples are referendums approved by voters such as Propositions M and Z in Santa Barbara and Monterey Counties, respectively. A proposed moratorium in Colorado would have restricted development drilling in certain counties. New York has effectively banned development of the Marcellus Shale even though the formation is a prolific producer in Pennsylvania. Evaluations of properties in areas subject to development and even continued operations should consider potential political risk but only in consultation with the operator and, when appropriate, relevant authorities.

## Environmental Offsets

When a project or proposed development activity may be located in an area where concerns about environmental impacts such as air and/or water pollution are a dominant issue, the governing authority may require that the pollution be offset by a reduction in air emissions at another location. Where such mitigation is not feasible, the purchase of "offsets" in the market place may be an option. Without getting into the complex world of environmental offsets, suffice it to say that the market is very tight and expensive. Absent the ability to obtain appropriate offsets, a project may be rejected by regulatory authorities.

# 13

# ABANDONMENT, RESTORATION, AND REMEDIATION

An increasingly important component of the operating and capital costs resulting from regulatory compliance is *Abandonment, Restoration,* and *Remediation* ("AR&R")[20]. Eventually, all oil & gas properties either reach physical depletion or, more commonly, and for a variety of reasons, become uneconomic for future operations. At that point, most oil & gas lease agreements, along with a growing body of state and local laws, require that the property be properly abandoned. **Abandonment** of the property includes plugging the wellbore to prevent fluid migration to the surface and/or between geologic intervals, removing all down-hole equipment such as pumps, tubing, etc., removing surface equipment including flow-lines, tanks, treating equipment, and dismantling of associated installations. **Restoration** is the return of the surface to its original contours if not condition. Finally, if necessary and as directed by one or more authorities, conduct **Remediation** of any soil and/or water contamination that may have occurred as the result of petroleum operations on the property.

AR&R is a legal obligation of the operator and a financial liability against the property. As production in many areas of the US has matured, and as environmental issues have gained greater leverage, the consideration of AR&R costs in oil property evaluation has received belated attention, both because of the imminence of those costs and the potential economic impact on the operator of the necessary expenditures. Physical abandonment of wells and facilities is well-documented activity; the engineering and mechanical aspects are not difficult. However, in order to properly determine the economic value of a property, the anticipated AR&R costs for that property must be accounted for as a reduction of income from the property prior to the scheduled abandonment. This is a relatively new concept to petroleum property evaluation and requires some elaboration.

## Past Practices

Common practices in the past toward treatment of AR&R costs fell into three categories: Ignore, Salvage, or Sale. Most evaluators simply ignored AR&R costs under the theory that the event was always years away and the company (owner) would provide the necessary funds when it came time to abandon the wells and properties. As a practical matter, any potential AR&R costs were (hopefully) a long time in the future, restoration was not considered a large issue, and the standards that existed for abandonment and, particularly, remediation were not strict and/or well-codified. However, as regulatory standards have expanded with increased attention to environmental concerns amid the desire in many areas for surface land development, the potential for near-term AR&R expenditures has begun to be recognized and to be included in valuation estimates.

Then there was the expectation that abandonment and restoration costs would be at least partially offset by anticipated salvage of well equipment (recoverable tubing, rods, pumping units) and facilities. However, the oil business is not a growth industry in many regions, and there may not be much market for old tubing and pumping units. Salvage companies often have to be paid to haul away tanks and other equipment as scrap. Further, the recognition that many wells and properties were old (and getting older), with outdated and deteriorated down-hole and surface equipment, meant that the salvage option often does not exist or is insufficient to offset any part of expected AR&R costs. Salvage value should not be considered an automatic offset. Any current salvage value that does exist will have to be depreciated to reflect expected value at the time of salvage.

The third method of treating AR&R cost is more subtle. The **Sale** (aka **Transfer of Responsibility**) approach assumed that the property would be sold or transferred to another operator before the liability would "come due"; the AR&R costs would then devolve on to the new operator. This is actually another form of the ignore syndrome. This approach no longer works very well as property investors have begun to recognize the AR&R liability when buying properties and, in many cases, have reduced the amount they are willing to pay or have written the Purchase and Sale (P&S) agreement to require the seller to fund the AR&R costs. Conversely, sellers of properties have also recognized the liability and have written P&S agreements that require the buyer to assume all or a large part of AR&R responsibility. Further, because many states have laws that allow the state to hold prior operators liable for AR&R costs on properties that they may have sold, in the event that the later operator defaults, sellers have often included terms in P&S documents that purchasers agree to specific conditions regarding abandonment, restoration, and remediation of purchased properties.

Transferring responsibility for AR&R to the buyer of a property can be done in any number of ways. The most direct (Buyer) approach is to accept responsibility, estimate the current cost of identifiable AR&R liabilities, and deduct that amount from the purchase price. Three other (Seller) methods incorporated into the purchase and sale agreement involve:

1. setting up a predetermined abandonment schedule,
2. establishing supervised escrow accounts to provide future funding, or
3. a combination of predetermined schedules and escrowed funding.

In a predetermined abandonment schedule, the new owner/operator is required to identify and abandon a certain number of (usually) idle wells each year with the objective that all "idle" wells, usually based on the prevailing regulatory definition of "idle," be abandoned over an agreed-upon number of years or by a certain future date. In some cases, specific wells and the year of abandonment are listed. The escrow account method requires the new operator to set aside a certain amount of funds each year into a managed escrow account to be used to meet eventual AR&R costs, often with a legally enforceable stipulation that a target number of wells will be abandoned. The escrow account approach gives more control to the Buyer. Both approaches tend to attract litigation, predominantly by Sellers attempting to enforce P&S provisions.

# Regulatory Necessity

Abandonment is not a new regulatory issue, and procedures are often codified. State and local governments have become more aware of the potential large obligation related to the AR&R aspects of oil & gas operations. A source of the concern is the increase in population in many regions and the resulting demand for expanded surface land development for housing, etc. Oil & gas fields and operations that used to be many miles from the nearest residential, commercial, or agricultural development are increasingly being encroached upon through demands by surface owners for access to build homes or use the surface for intensive agriculture. While not necessarily requiring abandonment of all operations, the increased surface use often demands that properties be cleaned up and consolidated which often includes the abandonment of idle wells and restoration of surface areas.

Further, certain states, such as California, have recognized that there are literally thousands of idle wells in the state that should have been abandoned according to state regulations but which remain in existence due to limited enforcement of regulations and a lack of incentive for operators to incur the expense of abandonment and restoration. A recent (2018) report by the California Department of Conservation stated that there were approximately 35,000 wells classed as idle in the state.[21] The state has also found that the historic practice of requiring good faith bonds to be deposited with the state is simply not sufficient to cover the costs for either individual wells or groups of wells. California now requires that the operator of wells that have been idle for two or more years and have not yet been properly abandoned enter into an Idle Well Management Plan which schedules future abandonments. The plans require that idle wells be physically abandoned and the well sites restored, thus forcing operators to recognize the AR&R costs as a near-term expenditure. The argument is often framed in terms of concerns (threats) to ground water and public safety, which is commendable, but the greater incentive to the state is to avoid the potential for the abandonment to become a state liability, particularly as many of the wells/properties are managed by companies and individuals that may not be sufficiently financed to undertake an AR&R program at a later date.

*"In California, an idle well is a well that has not been used for two years or more and has not yet been properly plugged and abandoned (sealed and closed). Plugging and abandonment involves permanently sealing the well with a cement plug to isolate the hydrocarbon-bearing formation from water sources and prevent leakage to the surface. If a well is not properly sealed and closed, it may provide a pathway for hydrocarbons or other contaminants to migrate into drinking water or to the surface."*[22]

Further, *"If an operator becomes insolvent or deserts their idle wells, responsibility for permanently sealing and closing these wells may fall to the State....To reduce the number of idle wells for which the state may become responsible, legislative and regulatory changes have been made to create incentives for operators to manage and eliminate their idle wells by entering into Idle Well Management Plans (IWMPs). If an operator does not have an IWMP, the annual idle well fees ...have been increased to reflect the potential costs associated with those wells. The fees are deposited into the Hazardous and Idle-Deserted Well Abatement fund to help fund the permanent sealing and closure of deserted wells."*[23]

Operators in California are assessed "idle well fees" annually based on the number of idle wells which existed at any time in the prior calendar year.[24] Failure to file the required idle well fee is taken to be conclusive evidence of desertion which then permits the State Supervisor of Oil and Gas to order the well plugged and abandoned. The regulations go on to require fluid-level testing and casing-pressure tests in wells that are idle. In reality, while the state can require abandonment of idle wells, it is not uncommon for companies to operate a well for a period of time to avoid the idle status. Many companies facing the AR&R requirement are not sufficiently financed to undertake a comprehensive AR&R program; the abandonment requirement and actual performance get bounced back and forth until (usually) the operator defaults. In that circumstance, the state will usually pursue any previous operators of the property, whether they drilled the wells or not, and, in some circumstances, the mineral rights owner.

When oil & gas operations exist in urban and suburban areas, particularly where the surface is planned to be used for residential or commercial development, it is not unusual for cities and/or counties to require testing and re-abandonment of wells which may have been abandoned many years before. While many such requirements exist, the City of Huntington Beach serves as a good example. New regulations came about in Huntington Beach primarily as the result of severe problems with gas (methane) leakage to the surface from old, shallow production wells in the Town Lot area of the Huntington Beach oil field underlying the City of Huntington Beach. Many of the wells had been previously abandoned, at least by the standards of the 1940s and 50s; however, the increase in real estate values in the late 1980s led to development of townhomes on the sites of many of the old wells. This would not have been a problem except that there was still active production, including waterflooding, being conducted in deeper horizons of the field. Migration of fluids (both liquids and gas) resulted in increased reservoir pressures in shallow intervals which caused many of the old shallow wells to leak methane to the surface. The leaks were found to result in more than one fire in new homes. The city imposed new regulations on real estate developers (the original oil operators are long gone) that required digging up the old wellheads (even those that were cut off), re-plugging the surface string, and welding a specifically designed plate on top of the well. In some cases, the developer had to install a gravel collar around the cut-off wellhead with a vent pipe extending to the surface.

## Securities and Exchange Commission Requirement

Another part of the regulatory pressure to define and incorporate AR&R costs in evaluations is provided in federal regulations for publicly traded companies. The regulations of the Securities and Exchange Commission (SEC) regarding financial reporting are intended to create financial statements that allow comparison among oil & gas companies. The SEC standards are not intended to be used as criteria for economic evaluations; there can be debate on this point, but that issue is outside the scope of this discussion. However, the SEC position on certain aspects of oil & gas operations has resulted in changes in both company financial reporting and in evaluation practices.

According to SEC regulations (see FASB #143)[25], an oil & gas company must determine any future liability resulting from AR&R of wells and other facilities. This is referred to as an **Asset Retirement Obligation (ARO)** which is a legal and financial obligation associated with the retirement of a tangible long-lived asset, in which the timing or method of settlement may be conditional on a future event the occurrence of which may not be within the control of the entity burdened by the obligation. ARO accounting is particularly for remediation work needed to restore a property, such as removing underground storage tanks, cleanup around an oil well, or removal of improvements to a site.

The potential costs of that liability must be estimated and a sufficient amount sent aside to fund those future costs. The method of accumulation is not specified. The response of companies to this requirement varies but, in many cases, the estimated abandonment and restoration costs have been made a part of the budgeting process down to the individual field and property level. In other words, the SEC made AR&R a subject of present evaluation concern.

## Real Estate Appraisal Practice

As the need to recognize and account for future AR&R costs in oil property evaluation has become more obvious, some of the methods used in real estate appraisal have been imported into oil & gas (mineral rights) evaluation. This has been driven largely by the need to evaluate properties for market value purposes, such as property tax, but real estate evaluation is not unique and there is no reason that methods for accumulation of funds to cover future expenditures should not apply to oil property evaluation.

There is a common practice in real estate appraisal known as a **replacement allowance.** Using this approach, the evaluator estimates the funds necessary for predictable major expenditures – a new roof on an apartment building or replacement of the HVAC system in an office building – and then sets up a schedule for the accumulation of the needed funds from the anticipated income stream for the apartment/office building.

The *Appraisal of Real Estate* notes, "*The...allowance for each component (of replacement) is usually estimated as the anticipated cost of its replacement prorated over its anticipated remaining economic life.*"[26]

> As an example, assume that a 40-year-old office building (such as the one in which I work) has an HVAC system that is 10 years old and has a remaining life of 10 years. A replacement unit costs $50,000. If rents from the building result in net cash flow of $2,000 per month the owner could set aside $500 per month deposited at a safe rate of 3% and accumulate the replacement cost over the remaining 10 years of life. By extension, a group of wells that can reasonably be scheduled for abandonment in 10 years can use a similar approach to accumulate funds from cash flow.

## Estimating AR&R Costs

Potential AR&R costs impact the value of oil properties. A buyer is not likely to pay full price for a property if there is the potential for substantial AR&R expense during and/or at the end of the remaining economic life of the property. Similarly, a seller may be willing to accept a lower price for a property if he can legally (and reliably) shift responsibility for future AR&R to the buyer. In recent years, purchase and sale agreements for oil property acquisitions have contained some especially creative language and procedures for transferring legal and financial responsibility for future abandonment and remediation from the seller to the buyer.

Once the potential AR&R liability is identified, an evaluator must then develop a schedule for abandoning all existing production and injection wells, idle and active, along with all wells anticipated to be drilled in the future. The cost of abandonment of all the wells plus restoration of the surface must be estimated. Well abandonment costs are a function of the depth and age of the well, the complexity of the completion interval, and the extent of the regulations. It should be expected that the costs of abandonment and restoration will increase over time, while the value of any salvageable equipment will decline.

Contingencies and complications should be anticipated. Old wells may, and probably do, have casing cement of questionable integrity, or the well may have been completed at a time when drilling and completion regulations and reporting requirements were less rigorous than those currently in effect. These wells are particularly troublesome where there is potential contact with, and possible contamination of, ground water. In areas that have been developed over decades, it is not unusual for an operator to have to re-abandon old wells to prevent communication with ground water or leakage to the surface. A modicum of uncertainty would be prudent. It may be possible to abandon 9 out of 10 wells at the estimated cost, but there is always the well with parted casing that has moved 100 feet away due to land slip. If environmental remediation is considered likely, that work must also be scheduled and a separate cost estimate completed.

## Sources of AR&R Data

The cost of abandonment of wells and facilities can often be estimated with reasonable accuracy based on the existing and foreseeable requirements of state and local agencies, particularly if the current operator and/or nearby operators have been actively abandoning wells. State and local agencies may also make data available. In the absence of either, an abandonment cost analysis may be necessary either through internal staff or a third-party contractor. Common practice is to schedule a number of wells to be abandoned at one time and retain a contractor to do the work.[27]

The cost of restoration of the surface will depend on local requirements, the nature and topography of the land, the intended future use, and existing regulations. Barring major unforeseen events, a good estimate should be possible with the help of qualified engineers, landscape planners, and, as necessary, local environmental experts. [Regarding the latter, it may be better to get them on board up front.] While unforeseen events are inherently hard to predict, the costs of the restoration portion of the AR&R process should not be difficult to ascertain. For the most part, restoration goes along with abandonment, so cost data should be available from documented operations, internal sources, and/or third-party consultants.

Remediation can be a major but often unquantified component of AR&R. Soil remediation may be necessary when it is found that tanks, flowlines, and/or wellheads have been leaking or where there have been unlined sumps in use. In old oil fields with tanks and sumps on every lease, it may be prudent to include some consideration of soil remediation. Many years ago (in the 1990s), a major oil company that owned a fee interest (surface and mineral rights) built a golf course among and around a large number of well and facility sites in northern Orange County, California. At some point management decided that some of the tank batteries and other surface facilities should be removed. In the process of removal, it was found that the bottoms of some of the tanks had deteriorated allowing crude oil to leak into the ground and subsequently spread out from the facility sites in the form of underground "plumes." The company spent several years trying to locate the plumes and then excavate and clean up the soil.

A particularly difficult problem occurs on properties where diluent (kerosene or other light oil) has been used to improve or enhance production of low-gravity oil. Unlike crude oil which tends to be viscous and, even when spilled on the surface, does not penetrate deep in the ground, kerosene and other light oils do tend to penetrate soils and can get into water supply formations. An example is the extensive litigation involving the Guadalupe oil field in San Luis Obispo County, California.[28]

The real difficulty with estimating remediation costs is that operators do not want to talk about potential remediation because that may suggest that it is necessary, and partly because of that, the costs are difficult to estimate because if an operator does the tests to determine if remediation is necessary, the operator may have to report the results to a government agency, which could (and likely will) trigger a requirement to do the work. Also, while a fair amount of experience with remediation costs has been accumulated by individual companies, a generally available database is nonexistent.[29] If there is reason to think there will be a need for remediation, a reasonable estimate should be made and included in the evaluation.[30] The real estate appraisal business has made some progress on this issue that may be applicable.[31]

Some caution is required because remediation costs can be high. Estimation of remediation costs may require one or more third-party studies and more than one cost estimate. The nature of most environmental regulation is such that once the work begins, it must continue until the work is determined to be completed by all parties concerned. In many jurisdictions, the process is considered to begin whenever an assessment study is started. If you are lucky, the contamination will not have spread far and is all on your property – but it can get out of hand. Groundwater contamination due to leakage or spills generally goes far beyond the actual costs of remediation to include litigation and related expenses.

Remediation analysis is property-specific as to the regulations in effect, the extent of the problem(s), and the potential use of the site. Where the only use for the next 100 years is for livestock grazing, soil remediation may not be as great a concern as it would be in an urban area where homes and schools are planned. Nonetheless, for each property being evaluated, the evaluator must consider and include the costs for potential remediation.

There is also an issue of managing properties with old abandoned wells and facilities, particularly where there is development of the surface for other purposes. Oil properties that are in the path of urban/suburban development present an interesting case. In many urban/suburban areas, surface developers have been required to locate and re-abandon wells and remove buried facilities such as pipelines before building homes or commercial structures. The purpose of this is to avoid problems caused by oil, gas, and/or water coming to the surface and damaging structures. This is particularly true where old, shallow wells exist and/or where oil & gas operations, which may communicate with shallower zones, are continuing. The cost of location and re-abandonment can be more than the original abandonment cost.

> As an example, in southern California, many old fields in Los Angeles, Ventura, and Orange Counties, that were developed in the early 1900s in open agricultural or ranch land, are now engulfed by expanding residential and commercial development. Even though large parts of these fields have been abandoned, portions are still being produced. The surface real estate is often worth more than the underlying oil production. In a few instances, an operator who owned the property in fee (as companies such as Chevron and Unocal did) consolidated their oil operations, abandoned outlying well pads, transferred the properties into separate entities, and built homes and commercial structures on the surface.

One excellent example is in Huntington Beach where luxury homes surround a golf course that in turn was built in and among the consolidated oil operations. Where the operator does not own the surface, it is not uncommon to relinquish the oil rights to a surface developer in exchange for the developer accepting the AR&R responsibility. There is an obvious advantage in this to the oil company, but the surface developer also benefits from the knowledge that wells are abandoned to current requirements.

## Applying AR&R Costs

The most direct, efficient, and logical application of AR&R costs for the property is as a direct offset to the estimated present value of the income stream for the property. The economic benefit of an oil & gas property is a function of the present value of the anticipated income stream, calculated using an appropriate discount rate: that is, the value on the effective date of the future income from the property. The present value of the AR&R costs is already known; it is the amount of the current estimate of those costs. Therefore, the value of the property can be estimated by subtracting the current estimate of AR&R costs from the present value of the income stream.

When the imposition of the abandonment and restoration is deferred, for whatever reason, either to some later point in the economic life or to the end of the economic life, the method of accumulation of the estimated cost becomes an issue. There are several ways to account for AR&R costs depending, to some extent, on the purpose of the evaluation. For a company's internal evaluation, it may be sufficient to say that abandonment costs will be paid by funds accumulated at the corporate level out of general revenues (as required by the SEC, for example). In this case, the costs show up on the balance sheet as ARO and are discussed in the financial notes. Where the company does so (and according to annual reports and 10-Ks, many do), this approach may be acceptable as long as the evaluator is satisfied that the estimated costs are reasonable and that there is adequate provision for accumulation. There is no reason that the ARO approach could not apply to non-public companies.

On the other hand, where the evaluation of a specific property is at issue, such as for ad valorem and other taxes or for eminent domain, the AR&R funds should be accumulated out of the income stream of the property being valued. The reason for accumulating funds out of projected future income is simply that at the time of abandonment, by definition, there is no income to pay for AR&R. An external or third-party source of financing should not be assumed. The evaluator should make an estimate of the costs and then devise a process for accumulating those funds (see Replacement Allowance). The scheduling of abandonment and restoration can be done in several ways, which, barring regulations or contractual requirements to the contrary, are up to the judgement of the evaluator. The general approach is to deduct a certain amount annually from the income stream that will add up to the projected AR&R cost.

Assume a property with an income stream from declining production that reaches economic limit in 10 years. The AR&R allowance could be accumulated in a number of ways:

1. a fixed amount as $/month or $/year,
2. a variable amount based on a percentage of revenue, or
3. a variable amount based on units of production ($/BO or $/MCF), or
4. a combination of methods.

The longer the life and the smaller the estimated AR&R cost, the less the impact on value. The approach is logical as well as good evaluation practice and is starting to show up in acquisition and internal company planning evaluations.

## Recognized by Regulatory Authorities

In regard to abandonment expenses for oil properties, the California State Board of Equalization (CSBE) states, "*It should not be assumed that funds will be available for abandonment costs from other sources. Abandonment costs are specific to each property and should be paid for out of each property's income stream. The allocation should be enough so that the full costs of abandonment is available when the property has reached its economic limit. The evaluator should not make the assumption that abandonment costs will occur after the economic limit has been reached.*" The funds may be accumulated through a simple deduction from income stream or "*Abandonment expenses may be accounted for by establishing a sinking fund account. The purpose of this account is to set aside funds each year. These funds grow to a predetermined amount by the time they are needed to pay for the cost of abandoning the property."*[32]

## Application of the Hoskold Method

The use of a sinking fund to aid in accumulating the estimated amount can be useful. For evaluation purposes, the sinking fund format, known as the Hoskold approach, is both sophisticated and appropriate. There are other methods, but Hoskold was developed specifically for application to wasting assets (coal mines) and is directly applicable to oil & gas properties.

As an example:

1. Assume an oil-producing property with 100 wells of which 75 wells are Producing and expected to remain on production for the remaining economic life of the property and 25 wells are Idle and scheduled for abandonment.
2. Assume that the Idle wells will be abandoned at the rate of 5 per year over the first five years and that the cost will be taken from operating cash flow.
3. Assume that oil production is declining at 4% per year, that oil prices are projected to be flat at $60 per barrel, and that all operating costs and deductions total $9,000,000 per year.
4. Assume that all remaining producing wells will be abandoned at the end of life in year 30.
5. Assume that a study has found that current abandonment and restoration costs for each of the remaining 75 wells is $300,000 and that these costs are anticipated to increase at an average of 3% per year. The study also anticipates that Remediation costs in Year 30 will be $12,000,000. The total cost is estimated to be $65,812,188.
6. Finally, assume that the anticipated AR&R costs are to be accumulated by setting aside 1% of Net Cash Flow (after operating costs and Idle well abandonment) into an account bearing interest at 4.55% per year.

As shown in the spreadsheet in Table 2. the property can accumulate an amount ($65,967,585) sufficient to cover most, if not all, the anticipated AR&R costs from the rather small set-aside amount through compound interest at 4.55%. The example is a relatively simplistic and somewhat static calculation that could be adjusted as time goes along and conditions change. It is likely that rather than wait until end of economic life to begin abandonments, the operator would begin to abandon groups of low-volume or poor performing wells earlier in the life of the property. Or, the operator could adjust the set-aside provisions to ensure that the expected amount is actually recovered out of production and sales, with perhaps a surplus accumulation to cover any surprises.

| Production Year | Oil Production Bbls | Oil Price $/Bbl | Oil Income $ | Total Operating Costs and Deductions $ | Net Operating Income $ | Producing Wells | Idle Wells | N | Inflation Factor 3% | Idle Well AR&R Cost $ | Net Cash Flow $ | Projected AR&R Cost $ | AR&R Cost Set Aside $ | Safe Interest 4.55% | Accumulated AR&R Cost Set Aside $ | Accumulated AR&R Cost Set Aside $ | Accumulated AR&R Cost Set Aside $ | Accumulated AR&R Cost Set Aside $ | Accumulated AR&R Cost Set Aside $ | Accumulated AR&R Cost Set Aside $ | Accumulated AR&R Cost Set Aside $ | Accumulated AR&R Cost Set Aside $ | Accumulated AR&R Cost Set Aside $ | Accumulated AR&R Cost Set Aside $ |
|---|---|---|---|---|---|---|---|---|---|---|---|---|---|---|---|---|---|---|---|---|---|---|---|---|
| 1 | 500,000 | 60.00 | 30,000,000 | 9,000,000 | 21,000,000 | 75 | 25 | 0.50 | 1.0000 | 1,500,000 | 19,500,000 | | 2,925,000 | 1.0000 | 2,925,000 | | | | | | | | | 2,925,000 |
| 2 | 480,000 | 60.00 | 28,800,000 | 9,000,000 | 19,800,000 | 75 | 20 | 1.50 | 1.0453 | 1,568,004 | 18,231,996 | | 2,734,799 | 1.0455 | 3,058,088 | 2,734,799 | | | | | | | | 5,792,887 |
| 3 | 460,800 | 60.00 | 27,648,000 | 9,000,000 | 18,648,000 | 75 | 15 | 2.50 | 1.0767 | 1,615,044 | 17,032,956 | | 2,554,943 | 1.0455 | 3,197,230 | 2,859,233 | 2,554,943 | | | | | | | 8,611,407 |
| 4 | 442,368 | 60.00 | 26,542,080 | 9,000,000 | 17,542,080 | 75 | 10 | 3.50 | 1.1090 | 1,663,495 | 15,878,585 | | 2,381,788 | 1.0455 | 3,342,704 | 2,989,328 | 2,671,193 | 2,381,788 | | | | | | 11,385,013 |
| 5 | 424,673 | 60.00 | 25,480,397 | 9,000,000 | 16,480,397 | 75 | 5 | 4.50 | 1.1423 | 1,713,400 | 14,766,997 | | 2,215,050 | 1.0455 | 3,494,798 | 3,125,342 | 2,792,733 | 2,490,159 | 2,215,050 | | | | | 14,118,081 |
| 6 | 407,686 | 60.00 | 24,461,181 | 9,000,000 | 15,461,181 | 75 | | 5.50 | 1.1765 | | 15,461,181 | | 2,319,177 | 1.0455 | 3,653,811 | 3,267,545 | 2,919,802 | 2,603,461 | 2,315,834 | 2,319,177 | | | | 17,079,631 |
| 7 | 391,379 | 60.00 | 23,482,734 | 9,000,000 | 14,482,734 | 75 | | 6.50 | 1.2118 | | 14,482,734 | | 2,172,410 | 1.0455 | 3,820,059 | 3,416,219 | 3,052,653 | 2,721,919 | 2,421,205 | 2,424,700 | 2,172,410 | | | 20,029,164 |
| 8 | 375,724 | 60.00 | 22,543,424 | 9,000,000 | 13,543,424 | 75 | | 7.50 | 1.2482 | | 13,543,424 | | 2,031,514 | 1.0455 | 3,993,872 | 3,571,657 | 3,191,549 | 2,845,766 | 2,531,370 | 2,535,024 | 2,271,255 | 2,031,514 | | 22,972,005 |
| 9 | 360,695 | 60.00 | 21,641,687 | 9,000,000 | 12,641,687 | 75 | | 8.50 | 1.2856 | | 12,641,687 | | 1,896,253 | 1.0455 | 4,175,593 | 3,734,167 | 3,336,764 | 2,975,248 | 2,646,547 | 2,650,367 | 2,374,597 | 2,123,948 | 1,896,253 | 25,913,484 |
| 10 | 346,267 | 60.00 | 20,776,020 | 9,000,000 | 11,776,020 | 75 | | 9.50 | 1.3242 | | 11,776,020 | | | 1.0455 | 4,365,583 | 3,904,072 | 3,488,587 | 3,110,622 | 2,766,965 | 2,770,959 | 2,482,641 | 2,220,587 | 1,982,533 | 27,092,548 |
| 11 | 332,416 | 60.00 | 19,944,979 | 9,000,000 | 10,944,979 | 75 | | 10.50 | 1.3639 | | 10,944,979 | | | 1.0455 | 4,564,217 | 4,081,707 | 3,647,318 | 3,252,156 | 2,892,862 | 2,897,037 | 2,595,601 | 2,321,624 | 2,072,738 | 28,325,259 |
| 12 | 319,120 | 60.00 | 19,147,180 | 9,000,000 | 10,147,180 | 75 | | 11.50 | 1.4048 | | 10,147,180 | | | 1.0455 | 4,771,888 | 4,267,425 | 3,813,271 | 3,400,129 | 3,024,487 | 3,028,853 | 2,713,701 | 2,427,258 | 2,167,047 | 29,614,058 |
| 13 | 306,355 | 60.00 | 18,381,293 | 9,000,000 | 9,381,293 | 75 | | 12.50 | 1.4470 | | 9,381,293 | | | 1.0455 | 4,989,009 | 4,461,592 | 3,986,774 | 3,554,835 | 3,162,101 | 3,166,665 | 2,837,174 | 2,537,698 | 2,265,648 | 30,961,497 |
| 14 | 294,101 | 60.00 | 17,646,041 | 9,000,000 | 8,646,041 | 75 | | 13.50 | 1.4904 | | 8,646,041 | | | 1.0455 | 5,216,009 | 4,664,595 | 4,168,173 | 3,716,579 | 3,305,977 | 3,310,749 | 2,966,266 | 2,653,163 | 2,368,735 | 32,370,246 |
| 15 | 282,337 | 60.00 | 16,940,199 | 9,000,000 | 7,940,199 | 75 | | 14.50 | 1.5351 | | 7,940,199 | | | 1.0455 | 5,453,338 | 4,876,834 | 4,357,824 | 3,885,684 | 3,456,399 | 3,461,388 | 3,101,231 | 2,773,882 | 2,476,513 | 33,843,092 |
| 16 | 271,043 | 60.00 | 16,262,591 | 9,000,000 | 7,262,591 | 75 | | 15.50 | 1.5812 | | 7,262,591 | | | 1.0455 | 5,701,465 | 5,098,730 | 4,556,105 | 4,062,482 | 3,613,665 | 3,618,881 | 3,242,337 | 2,900,094 | 2,589,194 | 35,382,952 |
| 17 | 260,201 | 60.00 | 15,612,088 | 9,000,000 | 6,612,088 | 75 | | 16.50 | 1.6286 | | 6,612,088 | | | 1.0455 | 5,960,881 | 5,330,722 | 4,763,408 | 4,247,325 | 3,778,086 | 3,783,540 | 3,389,863 | 3,032,048 | 2,707,002 | 36,992,877 |
| 18 | 249,793 | 60.00 | 14,987,604 | 9,000,000 | 5,987,604 | 75 | | 17.50 | 1.6775 | | 5,987,604 | | | 1.0455 | 6,232,101 | 5,573,270 | 4,980,143 | 4,440,579 | 3,949,989 | 3,955,691 | 3,544,102 | 3,170,006 | 2,830,171 | 38,676,053 |
| 19 | 239,802 | 60.00 | 14,388,100 | 9,000,000 | 5,388,100 | 75 | | 18.50 | 1.7278 | | 5,388,100 | | | 1.0455 | 6,515,662 | 5,826,854 | 5,206,740 | 4,642,625 | 4,129,714 | 4,135,675 | 3,705,359 | 3,314,242 | 2,958,944 | 40,435,813 |
| 20 | 230,210 | 60.00 | 13,812,576 | 9,000,000 | 4,812,576 | 75 | | 19.50 | 1.7796 | | 4,812,576 | | | 1.0455 | 6,812,124 | 6,091,976 | 5,443,647 | 4,853,864 | 4,317,616 | 4,323,848 | 3,873,952 | 3,465,040 | 3,093,575 | 42,275,643 |
| 21 | 221,001 | 60.00 | 13,260,073 | 9,000,000 | 4,260,073 | 75 | | 20.50 | 1.8330 | | 4,260,073 | | | 1.0455 | 7,122,076 | 6,369,160 | 5,691,332 | 5,074,715 | 4,514,067 | 4,520,583 | 4,050,217 | 3,622,699 | 3,234,333 | 44,199,184 |
| 22 | 212,161 | 60.00 | 12,729,670 | 9,000,000 | 3,729,670 | 75 | | 21.50 | 1.8880 | | 3,729,670 | | | 1.0455 | 7,446,131 | 6,658,957 | 5,950,288 | 5,305,615 | 4,719,457 | 4,726,270 | 4,234,502 | 3,787,532 | 3,381,495 | 46,210,247 |
| 23 | 203,675 | 60.00 | 12,220,483 | 9,000,000 | 3,220,483 | 75 | | 22.50 | 1.9446 | | 3,220,483 | | | 1.0455 | 7,784,930 | 6,961,940 | 6,221,026 | 5,547,020 | 4,934,193 | 4,941,315 | 4,427,172 | 3,959,864 | 3,535,353 | 48,312,813 |
| 24 | 195,528 | 60.00 | 11,731,664 | 9,000,000 | 2,731,664 | 75 | | 23.50 | 2.0030 | | 2,731,664 | | | 1.0455 | 8,139,144 | 7,278,708 | 6,504,083 | 5,799,410 | 5,158,698 | 5,166,145 | 4,628,608 | 4,140,038 | 3,696,212 | 50,511,046 |
| 25 | 187,707 | 60.00 | 11,262,397 | 9,000,000 | 2,262,397 | 75 | | 24.50 | 2.0631 | | 2,262,397 | | | 1.0455 | 8,509,475 | 7,609,889 | 6,800,019 | 6,063,283 | 5,393,419 | 5,401,205 | 4,839,210 | 4,328,410 | 3,864,390 | 52,809,299 |
| 26 | 180,198 | 60.00 | 10,811,902 | 9,000,000 | 1,811,902 | 75 | | 25.50 | 2.1250 | | 1,811,902 | | | 1.0455 | 8,896,656 | 7,956,139 | 7,109,420 | 6,339,162 | 5,638,820 | 5,646,959 | 5,059,394 | 4,525,352 | 4,040,219 | 55,212,122 |
| 27 | 172,990 | 60.00 | 10,379,425 | 9,000,000 | 1,379,425 | 75 | | 26.50 | 2.1887 | | 1,379,425 | | | 1.0455 | 9,301,454 | 8,318,144 | 7,432,898 | 6,627,594 | 5,895,386 | 5,903,896 | 5,289,597 | 4,731,256 | 4,224,049 | 57,724,274 |
| 28 | 166,071 | 60.00 | 9,964,248 | 9,000,000 | 964,248 | 75 | | 27.50 | 2.2544 | | 964,248 | | | 1.0455 | 9,724,670 | 8,696,619 | 7,771,095 | 6,929,150 | 6,163,626 | 6,172,523 | 5,530,273 | 4,946,528 | 4,416,243 | 60,350,728 |
| 29 | 159,428 | 60.00 | 9,565,678 | 9,000,000 | 565,678 | 75 | | 28.50 | 2.3220 | | 565,678 | | | 1.0455 | 10,167,143 | 9,092,315 | 8,124,680 | 7,244,426 | 6,444,071 | 6,453,373 | 5,781,901 | 5,171,595 | 4,617,183 | 63,096,686 |
| 30 | 153,051 | 60.00 | 9,183,051 | 9,000,000 | 183,051 | 75 | | 29.50 | 2.3917 | | 183,051 | 65,812,188 | | 1.0455 | 10,629,747 | 9,506,016 | 8,494,353 | 7,574,047 | 6,737,276 | 6,747,002 | 6,044,977 | 5,406,903 | 4,827,264 | 65,967,585 |
| | 8,826,779 | | 529,606,768 | 270,000,000 | 259,606,768 | | | | | 8,059,943 | | 65,812,188 | 21,230,934 | | | | | | | | | | | |

**Table 2.** Accumulation of Anticipated Abandonment, Restoration and Remediation Costs

# 14

# DISCOUNTED CASH FLOW FOR PETROLEUM PROPERTY EVALUATION

## The Present Value of Future Income

The value of any income-producing asset can be estimated by evaluation of the "stream" of income generated by that asset. The income stream comprises all the elements of the evaluation that are unique to the property being evaluated in the production projection, estimates of future prices, operating costs, capital expenditures, and net income. The income stream is a projection of anticipated future income that, depending on the form and structure of the asset, may extend for years or decades. An evaluation may find that a property could be expected to provide total income of $10,000,000 over a period of 12 years. That may be a useful result for budgeting and other uses but it is not the value of the income stream.

Estimation of value requires a second element of valuation, the **Present Value Factor**, which reduces the future income stream to economic value: that is, the value today of that future stream of income. The present value factor is used to account for the opportunity cost associated with the deferral of income to sometime in the future or, more commonly, making an investment with the expectation of future returns. In real estate appraisal, the discount rate is commonly termed the *Capitalization Rate*. Most appraisal texts, regulatory documents, and legal opinions refer to capitalization rate or "*Cap Rate*." The use of this term (as opposed to discount rate) complicates the integration of real estate and oil property appraisal. However, terminology aside, the function is the same – to reduce future income to current or present value.

An income stream is a series of payments received over time. Assume an income stream consisting of equal amounts of $5,000 over 10 years. The cumulative income at the end of 10 years would be $50,000. However, from the perspective of the date of valuation, the $5,000 to be earned in years 2, 4, and 10 does not have the same value as the $5,000 to be earned in the first year. Because of potential inflation and other financial risks, the present value of income to be received in the future is reduced or discounted by an amount that varies with the evaluator's perceptions of the risks involved in deferring the income. Present value is the inverse of compounding interest where, instead of increasing the value of a currently held amount over time, the present value factor reduces, or discounts, future amounts to an equivalent value today. To carry the analogy further, for example, if a principal sum (*P*) is invested at an annual interest rate (*i*) for (*n*) years, the accumulated interest would increase (*P*) to a total future amount (*S*) as follows:

$$S = P(1+i)^N$$

On the other hand, if (*S*) is the amount of income to be received in year (N), and the anticipated return is (*i*), then the present value of (*S*) would be:

$$P = \frac{S}{(1+i)^N}, \text{ where N = 10}$$

The series of annual $5,000 payments are discounted to present value by setting each successive annual amount equal to (S) and setting (N) at the appropriate number of years. The selection of an appropriate N and *i* is very important. Income streams are rarely constant; they generally fluctuate over time, and are often calculated on a monthly basis. The basic equation then must be modified to reflect the conditions imposed on the evaluation. A mid-year discount is probably more common than year-end.

# The Income Approach and Discounted Cash Flow

In an earlier section there was a review of the three "Approaches to Value" which, borrowed from real estate appraisal, could have application to valuation of oil & gas properties. That review demonstrated that of the three methods, the Cost Approach, the Comparable Sales Approach, and the Income Approach, only the latter has any utility for the estimation of value of individual oil & gas properties.

The income approach has historically been the primary method of valuing oil properties, both within the industry and in the financial markets that deal with the oil industry. The published literature reaches back to the late 1800s (Hoskold). The most common early form is the **Payback** or **Payout** method, wherein the value of a property is estimated as a function of the income to be received from the property within a certain time period. The payout method requires an estimate of future income and can accommodate expected changes in production, prices, and costs. While often used as an adjunct to more sophisticated analysis, payout is a form of income approach and is the most commonly used non-discounted valuation approach.

With the passage of time, and particularly since the availability of computers, undiscounted methods have been supplanted by discounted income methods. The primary form of the Income Approach applicable to oil property valuation is the **Discounted Cash Flow (DCF)**. The DCF is a simple methodology. The actual organization and calculation of the DCF for a property can be time-consuming and complex, requiring data gathering and analysis, but the concept and application are not difficult. The method is particularly useful because the evaluator of a project or property can include within the calculation, or in multiple variations, virtually every definable physical, operational, and economic characteristic of the property.

A recent survey of professionals who specialize in oil & gas property evaluation reports that the DCF is the preferred valuation method of 85% of those surveyed and of an additional 8% as the second choice.[33] The overwhelming prevalence of and preference for the discounted cash flow as the method for evaluating oil properties stems from several useful attributes of the method. First, the DCF is, by definition, a projection of future income. Second, the method duplicates, as closely as possible, the form of expected operation and exploitation of the property being appraised. Oil properties generate income over a period of time, during which production, prices, costs, and other factors can be expected to change continually; a DCF evaluation can model those expectations. The DCF has the ability to account for virtually every conceivable event and variable that might impact the income stream over the anticipated life of the property. Third, the DCF method is exceptionally flexible in allowing for multiple outcomes to be tested, compared, and weighed by simple alteration of expected conditions. Fourth, the DCF construction can be broad and simple or detailed and complex, subject to the needs of the evaluation.

These attributes make the DCF a powerful, dynamic, and rather elegant method for estimating the future income and value of oil properties. Each of the components of the cash flow result from a mix of the objective and subjective – a combination of measurable fact and judgement based on experience. When constructed and applied to a specific property, the method can result in a DCF that is not only unique to that property but also to the effective date of the evaluation. Neither the Comparable sales nor the Cost approach have the ability to model future performance, and they provide none of the dexterity of the DCF.

There is one attribute of DCF that is often overlooked, if only because it is so obvious. If an evaluator analyzes a property using DCF, the income projection is a model specific to that property. The production rate is a function of the geology of the reservoir(s) that can be produced on that property – not a property in another field or state. It is also a function of the physical factors, fluid characteristics, reservoir mechanism, and operating system on that property – not the lease next door. The same is true of product prices: one does not use West Texas Intermediate (WTI) 23° API oil pricing to value a Kern River property producing 13° API crude and vice versa. If an appraisal is done for each of 10 properties using the Income Approach, the appraisal of each property would capture all the physical and economic characteristics that define that property and would convert those characteristics into a stream of income. The outcome would be 10 streams of dollars each with a (possibly) different size, shape, and duration. Only the income approach can accomplish this feat. All of the differences between the 10 properties are boiled down to 10 streams of future income. For value purposes, the source of that income is almost irrelevant. It remains only to convert those income streams to a present value – which brings us to the discount rate.

# The Discount Rate

As noted above but worthy of repetition, DCF has three directly related but distinct components: an income stream, a present value factor or discount rate, and a derived value.

## Income Stream

The purpose of the income stream income is to attempt to model the flow of income that may be anticipated from a given oil property (with specific geologic, reservoir, and production characteristics) under defined operating and economic conditions. The income stream can be an intricate model that produces an estimate of income that is *unique to the property* and *specific to the date of evaluation.* Every physical or economic factor is related to the others. A change in the input data or in an assumption used in one part of the income stream may affect some other component and require further change at another point in the model.

## Discount Rate

The second component of DCF is the discount rate. All discount rates are present value factors. The primary purpose of the discount rate is to reduce the expected Income Stream to a (present) Value. Some discount rates, if derived and selected correctly, can be used to estimate specific results, such as market value. The discount rate is not a function of the property being appraised; it is unrelated to production rates, oil gravity, location, well count, oil prices, or any other characteristic of the property, all of which is in the income stream. The discount rate for an oil property is a *financial* factor, which is directly related to and is a function of the financial return that is required by an investor in an oil property, which is, in turn, a direct function of the tolerance for risk of that investor relative to the other investors in the marketplace.

| Year | Income $ | 10% PWF | 10% PW Income $ | 15% PWF | 15% PW Income $ | 18% PWF | 18% PW Income $ |
|---|---|---|---|---|---|---|---|
| 1 | 100,000 | 0.95346 | 95,346 | 0.93250 | 93,250 | 0.92057 | 92,057 |
| 2 | 100,000 | 0.86678 | 86,678 | 0.81087 | 81,087 | 0.78015 | 78,015 |
| 3 | 100,000 | 0.78799 | 78,799 | 0.70511 | 70,511 | 0.66114 | 66,114 |
| 4 | 100,000 | 0.71635 | 71,635 | 0.61314 | 61,314 | 0.56029 | 56,029 |
| 5 | 100,000 | 0.65123 | 65,123 | 0.53316 | 53,316 | 0.47482 | 47,482 |
| 6 | 100,000 | 0.59203 | 59,203 | 0.46362 | 46,362 | 0.40239 | 40,239 |
| 7 | 100,000 | 0.53820 | 53,820 | 0.40315 | 40,315 | 0.34101 | 34,101 |
| 8 | 100,000 | 0.48928 | 48,928 | 0.35056 | 35,056 | 0.28899 | 28,899 |
| 9 | 100,000 | 0.44480 | 44,480 | 0.30484 | 30,484 | 0.24491 | 24,491 |
| 10 | 100,000 | 0.40436 | 40,436 | 0.26508 | 26,508 | 0.20755 | 20,755 |
| Total | 1,000,000 | | 644,448 | | 538,203 | | 488,182 |

**Table 3.** PWF is Present Worth Factor at Selected Discount Rate with Mid-Year Discounting

The selection and application of the discount rate is one of the most important elections made in evaluating a property or project. This is illustrated by comparing the present value of an income stream which provides $100,000 every year for 10 years. At increasing discount rate, the present value declines as shown in the table below. If, as in most oil & gas income streams, the cash flow declines rather than remaining flat, the effect of the discount rate is even more pronounced. Indeed, the discount rate can have the effect of reducing a well-constructed long-term production and income stream to modest present values.

## Value

After all the work of constructing a reliable income stream and selecting a discount rate, the evaluator puts the two together, discounts the cash flow, and arrives at the estimate of value – be it for internal purposes, purchase, sale, property or other taxes, eminent domain, civil damages, or abject curiosity. An evaluation that estimates value for any purpose and is a function of the income stream and the discount rate is unique to that property at the effective date. Assuming that the evaluator (investor) has put everything he knows of the property and his expectations for future economic conditions into the Income Stream and has applied an appropriate Discount Rate, then the Value estimate obtained should be correct for the purpose intended.

# Further Discussion of the Discount Rate – Misconceptions

The discount rate, as noted, can be a powerful element in property valuation, but the discount rate is also one of the least understood and poorly defined of any aspect of petroleum property valuation. Before going further, it would be useful to discuss, and then discard, certain misconceptions about the discount rate as it applies to oil property evaluations. Many of the misconceptions result from limited discussion within evaluation practice regarding the purpose, construction, and application of the discount rate and confusion within the discussion that does occur. Only a few petroleum evaluation texts, which may refer to DCF, include any discussion of either the source or application of appropriate discount rates. Despite the extensive body of financial literature regarding discount rates (often in excruciating detail), little of that information, with two exceptions to be discussed below, is translated into an oil & gas context. Also, while there are a number of SPE papers and other published industry sources that delve into sources and/or applications, those papers are limited in number and are often contradictory. Part of the misconception and confusion regarding appropriate discount rates is seen in those same publications and in industry surveys and studies that include discount rate discussion.

One common source of misunderstanding of the discount rate is the income tax issue. The discount rate for valuing oil & gas properties is the **Before Federal Income Tax (BFIT)** rate. The reason is simple – individual properties do not pay income tax. The owner/operator, be it corporation or individual, pays tax on income received by the person or company, but the property does not. Applying income tax issues on the income from a property imposes conditions that relate to the taxpayer (tax rate, depreciation, etc.), not the property. Even for the same taxpayer, certain tax provisions may apply to some properties and not to others. Then there are the various state tax systems that have to be considered. Individual properties are comparable on a BFIT basis but not on an After Federal Income Tax (AFIT) basis.

The discount rate is not a function of the current bank interest rate or the interest rate on funds that could be borrowed from another source. Using a bank interest rate to value a property or an investment in that property would suggest that the investor or owner of the property would be willing to accept the equivalent of bank interest, whatever it might be, as the return for making the investment. The use of a bank or borrower interest rate from any source would have the effect of overvaluing the property.

One of the more intensely (often heatedly) debated aspects of the selection and application of the discount rate within DCF valuation of oil properties is the inclusion of "risk" in the discount rate. Many writers of SPE papers and contributors to surveys of valuation methods inadvertently create a very confused discussion of the inclusion of risk in the discount rate and how that risk should be expressed. As will be discussed below in the derivation of discount rates, the **financial source** includes a systemic measure of risk based on the uncertainty related to the oil & gas industry in general. However, when the same financial derivation is amended to include specific adjustments, those do not include any risk related to the specific property being evaluated. That risk assessment is separate and is accounted for in various ways, depending on the form of the uncertainty, as part of the evaluation, not in the discount rate. Here again, the reason should be obvious: a value derived from a discounted cash flow of Property A using a risk-inclusive discount rate cannot be compared to the present value of Property B derived using a different risk-inclusive discount rate (see discussion of Risk Adjustment below).

## The Financial or Cost-of-Capital Source

The following is a summary of the financial source of the discount rate. See Appendix C for a more detailed discussion. Depending upon the purpose of the evaluation, the discount rate to be used may be provided by company guidelines, by the requirements of financial institutions, or by government stipulation. Absent these sources, the evaluator can rely on professional judgment and derive a discount rate from experience and/or personal knowledge. In either case, the foundation for discount rate selection in the context of evaluation of oil & gas properties rests, in whole or in part, on financial concepts and should be derived from financial sources.

Many oil & gas companies routinely define a discount rate or range of discount rates for use in project evaluation and capital budgeting. This discount rate is a function of company policy and, as shown by Boyle and Schenck[34] and also by Dougherty and Sarkar,[35] is based on an estimate by management of the company's weighted average cost-of-capital. Finance-based discount rates are applied to the estimate of income from new projects, capital investments, and acquisitions/sales and are structured to allow comparison of those projects, etc., usually as part of the Net Present Value (NPV) and/or Internal Rate of Return (IRR) methods of capital budgeting.

The **Cost-of-Capital (COC)** is a financial analysis term that refers to the cost to an investor or investors of the funds that are to be used to make an investment and is the financial standard used in industry for investment analysis. Space here does not allow an extensive treatment of this topic; there are many financial and valuation textbooks as well as published articles available that cover the subject in excruciating detail. As noted above, oil industry sources are limited, however; a background in the subject is important to understanding petroleum investment and evaluation practices.

The COC can be described as the price, in the form of a rate-of-return, that must be "paid" in some form to investors in order to obtain and retain the funds necessary for capital investment projects. Put another way, the COC is "*the expected rate of return that the market requires in order to attract funds to a particular investment.*"[36]

The COC is not the cost of borrowed funds alone, the cost of equity funds alone, or a function of prime rate. Furthermore, it should be a return (cost) consistent with the term of the intended investment. Oil industry investments, such as acquisitions of producing properties, development of new production, or EOR projects, are relatively long-term investments requiring several years to pay out the initial investment and often much longer to reap the full benefit of the investment. The appropriate return is a function of 10- to 20-year debt instruments and equity held for the duration.

## Weighted Average Cost-of-Capital

A knowledge of the COC concept is essential to deriving appropriate discount rates for oil property evaluation. Many oil companies routinely define a range of discount rates for use in project evaluation and capital budgeting. This discount rate, as shown by several authors, is commonly based on an estimate of the company's **Weighted Average Cost-of-Capital (WACC)**. Finance-based discount rates are applied to the estimate of income from new projects, capital investments, and acquisitions/sales, and are structured to allow comparison of those projects, as part of the capital budgeting process.

If the derived cost-of-capital is to be used as a basis to value properties, it must be a prospective or forward rate. It is the combination of the respective returns anticipated by the holders of the debt and equity portions of the invested funds. In a small company, the debt holder might be the bank or other lender, while the equity holder(s) might be the individual investors or partners. In a large corporation, the debt return would be the yield expected by the public and institutional holders of the bonds and notes issued by the corporation, while the equity return would be the growth and income return anticipated by stockholders. In either case, it is the anticipated future return on debt and equity, not the historical return, that is at issue.

Because of the way it is constructed, the WACC is the COC to the company as a whole and represents the future expectations of debt and equity holders for returns on the company's operations. There is an implicit assumption that the company will continue to operate and finance projects in the same manner as before. The market "corrects" the value of the company's stock price and debt when expectations change or the company signals a change. The company COC "*is the correct discount rate for projects that have the same risk as the company's existing business but not for those projects that are safer or riskier than the company's average.*"[37]

# Adjustments to the Cost-of-Capital Discount Rate

The BFIT WACC can be used as the baseline from which to build a property- or project-specific discount rate. However, discount rates derived from WACC analysis cannot be applied directly to evaluation of individual oil properties. The WACC is deficient in several elements which must be considered before application can be made to a specific property. There must first be adjustments to account for inherent differences between WACC returns and the returns that would be expected from a specific property. The differences are several, complex, and important.

1. Natural resources such as oil & gas that diminish in value because of extraction of that resource are known as wasting assets. Oil & gas properties are wasting assets which are produced to depletion or economic limit. Petroleum companies are not considered to be wasting assets; companies expand, add product lines, and develop new properties.
2. The return derived from WACC analysis represents the return from a portfolio of debt and equity issued by public companies where reversion of the invested amount is presumed but not provided.
3. The WACC is the discount rate used to value the company, not the individual assets.
4. Individual oil property investments carry more risk than an investment in a corporation. An oil property is an investment dependent on one income stream derived from oil & gas production from that property. The components of WACC return represent returns from a wide portfolio of investments.

The return anticipated from a specific producing property must exceed the WACC to account for deficient areas which can include, but are not limited to, Liquidity, Return-of-Capital, and Portfolio Diversity.

## Liquidity

WACC is derived from the stocks and bonds of publicly traded companies. The Cost of Debt (COD) and Cost of Equity (COE) are the returns that are expected by investors in publicly traded debt and equity, respectively. These investments are highly liquid; they can be bought and sold almost instantaneously. The same cannot be said of individual oil properties. This illiquidity demands a premium on the cost-of-capital if it is to be used to value properties.

## Return-of-Investment

Investors generally require a return-of their investment and a return-on that investment (interest). The WACC, as commonly constructed, represents only a return-on-capital invested. The calculation of a WACC, at a specific time, assumes that the calculated return will continue from that time forward – in effect, a level and perpetual annuity (return-on-investment). The WACC assumes that the original investment is recovered when the debt or equity interest is sold. In finance, this is termed return-of-capital. In contrast, individual oil & gas properties have variable and declining income streams and are produced to depletion. There is a fixed volume of oil & gas; only a portion of the volume can be produced and sold. The production of part of the volume reduces the amount remaining to be produced. When all economically producible oil has been recovered, the asset will have been diminished to zero and will have no residual or reversionary value. This will be true even though changes in economic and technological conditions may act to defer the point of final depletion. There is no remaining asset value from which to recover the original investment. The income stream must provide both a return-on-investment and a return-of-investment, the combination of which must necessarily be greater than the WACC return alone.

## Portfolio Diversity

The largest difference between corporate cost-of-capital and the returns from investments in individual oil properties is the diversification of risk through a "Portfolio Effect." A company that owns a number of properties can expect an income stream from each, whereas a single property provides only one income stream. In the corporate context, there is greater possibility of loss in the single income stream than in the multiple streams, where a failure of one stream has a limited effect and may be offset by an increase in another income stream. The income stream from a particular property may have greater or lesser risk that the other income streams in the portfolio as the result of any number of engineering, geologic, operational, or economic factors.

# Discount Rate, Opportunity Cost, Hurdle Rate, and Minimum Required Return

Based on the above discussion and as amplified in Appendix C, it is clear that the financial basis of discount rates, either as corporate WACC or adjusted WACC, can be used to calculate the present value of a future stream of income from a particular property or project and, possibly, for a particular purpose. Further, the present value income stream from that property or project is directly comparable to the present value income stream from any other similar property or project. This is good insofar as it goes, but it raises the question of how the derived return should be used in evaluation of capital investments in general and oil & gas properties in particular.

According to commonly accepted financial theory as discussed in numerous texts and sources,[38,39,40,41,42] a company or other investor should attempt to make investments of capital only in projects that will provide a rate-of-return at least equal to the cost of the capital used to make the investment. However, these same texts and sources also offer a disturbing number of terms used to describe the rate-of-return that should be used. Without delving into every financial and appraisal text, the terms can be aggregated as discount rate, opportunity cost-of-capital, hurdle rate, or, to add a new term, **Minimum Required Return (MRR)**. The discount rate description is generic and imparts no new information.

One respected financial text, discussing the rate-of-return for an investment opportunity (project) states that, *"The rate of return is called the discount rate, hurdle rate, or opportunity cost of capital. It is the opportunity cost because it is the return that is foregone by investing in the project rather than investing in financial markets."*[43] Another equally valid source refers to "hurdle rate" with no differentiation to opportunity cost. Virtually all the other sources refer to and elaborate upon "opportunity cost-of-capital" while either ignoring hurdle rate and MRR or mentioning them only in passing. This situation is problematic for at least two reasons. First, the use of differing terms is confusing: Are the terms simply variations on a theme or are there functional differences that should be considered. Second, if the latter, how are we to know which variation to use for valuation?

As noted, most sources refer to the opportunity cost-of-capital and are relatively consistent in defining the term. In addition to the text source quoted above, several others sources are of interest. Taylor, writing at the dawn of time, notes regarding opportunity cost that, *"Every owner of capital has more than one opportunity to invest his money. Every time he accepts one of those opportunities he foregoes the opportunity of investing in another and, with it, foregoes the profit he could have made in that investment. This situation gives rise to the term opportunity cost. This concept holds that capital is never free, because selecting one use of capital implies the cost of foregoing the opportunity to earn a profit with it elsewhere."*[44]

Real estate appraisal offers another interesting approach, *"Opportunity cost is the net cost of opportunities not chosen or options foregone, denied, or lost. An investor who selects one investment foregoes the opportunity to invest in other available investments. An investor will select the investment that best meets his or her investment objectives....Opportunity cost is related to the principle of substitution, and is particularly significant in estimating the rates of return necessary to attract capital."*[45]

Alas, it is still not clear how to equate the opportunity cost concept to the evaluation of oil & gas properties. Financial sources allude to the idea that it is the *"return that is foregone by investing in the project rather than in financial markets."*[46] suggesting that the alternative to investing in a production project is to invest in stocks and bonds of companies similar to the investor or even in the company. But, we already know that corporate cost-of-capital does not account for liquidity, return-of-capital, and portfolio diversity; the discount rate for a single project income stream must necessarily be higher than the company cost-of-capital. This is where the transition to hurdle rate or MRR occurs and where the difference may become apparent.

Absent actual sources but working from experience and intuition, hurdle rate and MRR appear to be practical intrusions on the otherwise ordered world of business finance. To wit, *"Business people have good intuition about relative risks, at least in industries they are used to, but not about absolute risk or required rates of return. Therefore, they set a company- or industry-wide cost of capital as a benchmark. This is not the right hurdle rate for everything the company does, but judgmental adjustments can be made for more or less risky ventures."*[47]

The ever-practical Shannon Pratt goes a bit further, *"When a company uses the cost of capital to evaluate a commitment of capital to an investment or project, it often refers to that cost of capital as the "hurdle rate." By the "hurdle rate" they mean the minimum expected rate of return that they would be willing to accept to justify making the investment....the "hurdle rate" for any given prospective investment may be at, above, or below the company's overall cost of capital depending on the degree of risk of the prospective investment compared to the company's overall risk."*[48] This leads, necessarily, to the following: *"The most popular theme of contemporary corporate finance is that companies should be making investments, either capital investments or acquisitions, from which the returns will exceed the cost of capital for that investment. Doing so creates "economic value added," "economic profit," or "shareholder value added."*[49]

Campbell refers to the **Minimum Return Standard (MRS)** and notes, *"Every project is compared against a threshold to determine if it meets corporate objectives. Companies arrive at threshold value differently and, regardless of the company, a certain arbitrariness occurs in defining the standard."*[50] The WACC, or a variation of WACC, serves as the foundation for the "*hurdle rate*." In practice, the MRR or MRS may, and usually does, exceed the actual cost-of-capital by a few percentage points in order to provide a margin of return over the baseline WACC.

The result of this collection of experts is a tempered conclusion that either the terms (opportunity cost, discount rate, hurdle rate, or MRR) mean essentially the same thing, variations on a theme, or, as seems to be in practice, the opportunity cost, but more so, the hurdle rate and/or MRR, are defined by the investor relative to the company cost-of-capital. The WACC, or a variation of WACC, serves as the foundation for the hurdle rate or MRR representing a threshold value for investment decision making. In practice, the threshold rate may, and usually does, exceed the actual cost-of-capital by a few percentage points in order to provide a margin of return over the minimum.

> As an example, the BFIT WACC calculated for the sample of 19 companies compiled by the Texas Property Tax Assistance Division (PTAD) (see Appendix C, p. 263) of 12.45% in 2020. Using this data, a company could use a rounded-up rate of 12.50% as the cost-of-capital for the company. However,

the company recognizes that investing only to obtain the same return as the existing cost-of-capital does not necessarily act to attract capital or increase shareholder value. One way to do that is to require all capital investments to return an increment higher (say 2%) than the cost-of-capital (*minimum required return*). This does not mean that all the investments will actually return 14.5% but the potential is greater than if only 12.5% is used. As a practical example, the PTAD *"adds two percentage points to the overall mean WACC of 12.45% to establish a base discount rate of 14.45%."*[51]

## Some Thoughts on "Risk-Adjusting" the Discount Rate

It is not uncommon to find debate in industry literature and presentations about the practice of "adjusting" the discount rate (or hurdle rate or MRR, whichever term is used) to account for project risk. Indeed, there are surveys of property evaluation professionals that compare the use of risk-adjusted, often termed "risk-inclusive," discount rates to the use of non-adjusted discount rates. In essence, the **risk-adjustment** of the discount rate is done by increasing the discount rate to account for the perceived risk of the project. This is not a good idea.

As noted by Sinha and Poole, *"The practice of employing a higher return requirement for riskier projects such as exploration is not recommended. The selection of a higher return requirement is often arbitrary. This technique does not differentiate between the dissimilar risk profiles of each project within the same perceived risk category. By imposing [a] high return requirement on an ad hoc basis, a company may miss out on an investment opportunity which would have increased shareholder wealth. Project risk should be accounted for using Monte Carlo simulation or other statistical decision theories."*[52] In other words, the risks identified in the project should be quantified and accounted for in the construction of the discounted cash flow for the project, such as applying adjustments to the production projection and/or oil price projection prior to discounting rather than adding an arbitrary factor to the discount rate.

As an example, some practitioners have occasionally suggested that evaluations of California heavy oil projects should be subjected to a higher discount rate than, say, West Texas properties because of the alleged poor crude quality. It should not be, but often is, necessary to point out that the cash flows for those properties are constructed using the crude prices applied to, and operating costs incurred in, producing heavy oil and therefore a higher discount rate would be not just superfluous but a poor business decision.

The clarity expressed by Sinha and Poole would seem to be obvious and suggest that as the resources available to petroleum evaluation, including but not limited to computer-based data analysis systems, have advanced over the course of 35+ years, the practice of adding an increment to the discount rate to compensate for some form of risk or uncertainty in the evaluation would have been phased out. As noted above, the discount rate derived from the cost-of-capital for oil & gas companies already includes a measure of risk related to the petroleum industry as a whole. To the extent that the practice of adjusting discount rates for risk related to specific properties or projects continues, some discussion is necessary.

A relatively recent authoritative study provides some insight. The Society of Petroleum Evaluation Engineers (SPEE) has conducted an annual "Survey of Parameters Used in Property Evaluation" ("the Survey") for over 30 years. The composition of the Survey and the demographics of the population of respondents has changed with time but many aspects of the survey have been consistent although, in some cases, expanded. Along the way, the survey began to include questions regarding Risk Adjustment of the evaluation (or portions thereof) and Risk Adjustment of Discount Rates. The Survey report for 2014 states the following:

*"Most evaluations of oil and gas reserves are initially performed in an unrisked manner, with reserve categorizations... and reserves subdivisions... serving as the only indicators of relative risk, although these indicators are qualitative at best. In order to develop a value from unrisked DCF models, risk factors are usually applied. Two of the most common methods of incorporating risk into DCF models include risking the reserves (i.e., Risk Adjustment Factors) and risking the Discount Rate (i.e., use of Risk Adjusted Discount Rate).... When an evaluator uses increased discounting to account for risk in evaluating the... value of an oil/gas asset, that evaluator is using a Risk Adjusted Discount Rate* [RADR]. *The Risk Adjusted Discount Rate should include a consideration for the "cost of capital," a profit or expected rate of return... and any risk/uncertainty that the evaluator may choose to impute to the asset....if is fully expected that it will be larger than the firm's "cost of debt" or "cost of capital."* Further, Risk Adjustment Factors (RAFs) *"are used to downward-adjust reserves (or the cash flows derived there from) for risk by reserves status and category."*[53] The use of RAFs will be discussed in more detail in the section on Investment Analysis. It is sufficient at this point to note that 44% of 109 respondents use RAFs and do not risk-adjust the discount rate.

Despite some inconsistencies in definitions and an excessive reliance on "reserves" as the basis of evaluation, the Survey produces some interesting results. From the report, *"this Survey asked respondents to specify how they typically adjusted cash flow models to account for risk and uncertainty, given the choice of RADRs, RAFs, using a combination of the two methods, or using other methods."*[54] The other methods were not defined. According to the survey report, 24% of respondents (26) indicated that they used RAFs in conjunction with another discount rate adjustment method. Twenty-eight percent (28%) used RADRs exclusively.

Based on the Survey data reported, of the 57 respondents (52.3%) of all 109 respondents that said that they used RADRs exclusively or used RADRs in conjunction with RAFs, 91% said that they *"applied the RADR to cash flows before income tax."*[55] A smaller group (47 respondents) provided information about their typical RADR as well as other data. A group of 17 responses identified as relating to exploration and production (E&P) provided data that averaged 16.3% (range 13.9-27.2%), while a group of 23 Consultants provided responses that averaged 16.6% (range 12.8-27.2%). When the respondents were asked for the RADRs they applied to "proved developed producing" (undefined in the survey), the E&P group averaged 11.5% while the Consultant group reported an average of 11.1%.

*"The respondents were then asked to identify from a list of... risk factors, all those that influenced their decision regarding the specific RADR used."* The results of 54 responses are summarized in the table below.[56]

| Potential Risk Factors | Count of Responses | Percent of Respondents Incorporating Risk Factor into[57] RADR |
|---|---|---|
| Reserve Risk | 42 | 76 |
| Price Uncertainty | 36 | 65 |
| Expense Uncertainty | 33 | 60 |
| Mechanical Risk | 26 | 47 |
| Political/Regulatory Uncertainty | 18 | 33 |
| Profit | 18 | 33 |
| Unidentified Reserves Potential | 12 | 22 |
| Other | 2 | 4 |

**Table 4.** Potential risk factors ranking from SPEE Parameter Survey

The information provided in the 2014 SPEE Survey report represents the results of a survey taken in one year of a small but presumably sophisticated group of evaluation experts. A review of the Survey report for other years does not show differences from year to year which suggests that evaluation practices do not change very much with time and/or the annual responses are coming from a consistent pool of experts whose practices do not vary.

With that said, some useful information and points of discussion can be extracted from the Survey data. The following discussion uses the PTAD Study for 2014 and the SPEE Parameter Survey of the same year as sources of talking points and illustrations that could be extended to other years and circumstances.

## Comparison to Cost of Capital

As discussed in Appendix C, the Texas Study found an industry BFIT cost-of-capital of 13.46% for 2014.[58] The SPEE Survey reported an average RADR of 16.3-16.6% for the E&P and Consultant groups, respectively, with a relatively wide range. The comparison to the base COC is reasonable, allowing an increment of about 3% over the base rate. There is a difficulty, however, in that the Survey results are presented as an average of all RADRs, suggesting that the actual range of RADRs incorporates but does not necessarily exceed the COC. This is illustrated by the comparison of the RADR values for low-risk, Proved Developed Producing (PDP) properties, those with developed production, of 11.5% and 11.1%. PDP properties are likely to be the most commonly evaluated. Under the terms of the Survey results, the RADR used for such properties would be about 2% lower than the cost-of-capital for the same period. PDP properties may be low risk but they are not without risk. Using a RADR that is lower than the COC implies that the company is of greater risk than the property and that investors in PDP properties would accept a lower return than the same investors require from the company as a whole. While this may occur in certain circumstances where factors other than financial outcomes are in play, it cannot be a rational approach. There appears to be a dichotomy between understanding of cost-of-capital and risk analysis of properties.[59]

## Multiple Risk Elements

Half or greater of the respondents include two or more risk elements in the RADR with Reserve Risk, Price Uncertainty, and Expense Uncertainty being the leading concerns and Mechanical Risk close behind in fourth place. Aside from the issue of quantification of the various risk elements (see below), there is the problem of how the risk elements are apportioned. Is reserve risk or price uncertainty the dominant element? Is expense risk, possibly influenced by inflation, a partial function of price uncertainty, and if so, how do they overlap or cancel each other out? The problem with using an RADR that includes multiple risk issues is that the lump sum adjustment does not allow for correct application of the risk elements. This is particularly true where one of the elements such as reserve risk could be applied separately using an RAF rather than being lumped together with other possible lesser risks or uncertainties.

## Quantification of Risk Elements

There is another obvious contradiction in the use of RADRs with multiple risk elements. Unless the selection of the RADR is completely arbitrary or (more charitably) based on experience and knowledge, it has to be assumed that the decision to include reserve risk and price uncertainty in the RADR is based on discrete increments added together – say, reserve risk is worth 3% and price uncertainty is worth 1%, both of which are added to the base COC discount rate (15.5% in 2014) for a total of 19.5%. What if political/regulatory uncertainty, estimated to be 3%, is also added in, bringing the total discount rate to 22.5%? How were the risk elements determined? We already know that RAFs for reserve risk exist and could be used to cover that concern, thereby reducing the need for that element in the RADR. If there is a 1% concern about the price how was the 1% determined? Is the uncertainty in the starting price or the price projection? More likely the latter. If so, could not the adjustment be applied to the price projection(s) by Monte Carlo simulation?

Admittedly, political /regulatory risk is a real issue, but this is usually a "should we do it or not" business decision that is not compatible with the discount rate. If the project being evaluated is in Turkmenistan, Russia, or, worst case, Santa Barbara County, there are sources that can provide potential country risk factors and/or regulatory commentary that can be used to help make the investment decision rather than incorporating a necessarily imprecise risk element into the discount rate.

## Treats All Risk Equally as Part of DCF

When a RADR is applied to a projected cash flow, the effect is to treat all parts of the cash flow as having the same total risk regardless of the contribution of the various risk elements to the cash flow. That is, if reserve risk is assessed at 3% and price uncertainty at 2%, the entire cash flow is risk adjusted at 5% including some parts of the cash flow such as operating costs, production taxes, and capital investments that may not be subject to risk. In contrast, a cash flow where each part is risk adjusted (as needed) independently reflects the actual perceived risk related to that part. The cash flow can then be discounted at the COC or MMR as required by company policy.

## Lack of Comparability Among Value Results

As noted above, and discussed further in the section on Investment Analysis, one of the most attractive features of the discounted cash flow method of valuation is that the results of the evaluation are directly comparable to each other, thus allowing business decisions to be rendered based on reliable analysis. If RADRs are used to value multiple cash flows, the results are not comparable.

# 15

# INVESTMENT ANALYSIS

As noted at the beginning of our discussion of property valuation, one of the primary purposes of economic evaluation is for internal business activities such as planning and budgeting of capital expenditures and comparison of investment opportunities. The personal vignette about my presentation of a production restoration project to my boss in 1968 was just such a use of economic evaluation to justify a major expenditure of capital by my employer. As you may remember, the project was politely deferred by Mr. Smith because, per his review, the project returns did not meet Texaco's investment criteria. I do not recall what those criteria were and, given the approval structure in place, I may not have known at the time. Suffice it to say that the company had certain criteria for considering production investments.

There is, of course, nothing unusual about the establishment of certain minimum return conditions for planning and budgeting for capital expenditures. Petroleum industry corporations, investors (shareholders) in those companies, and individuals who invest directly in oil & gas properties/projects have a desire to maximize wealth and to maintain overall value. To that end there are many methods available to estimate the potential value of investment opportunities and to compare the return on those investments so as to make the best use of available capital. It is important to note that the methods used for oil & gas investments are not unique; they are based on, derived from, and articulated in common business/financial texts and other sources, some of which will be introduced below.

## Industry Methods of Investment Analysis

Methods of measuring profitability and allocating funds to competing investments has been and remains a topic of discussion in business, financial, and (in some) petroleum analysis texts and papers. In 1985, Boyle and Schenck published a study of investment evaluation methods used in the oil & gas industry.[60] The study found that most companies allocate capital to investments in a systematic manner and tend to use certain basic methods of investment analysis. The study consisted of analysis of data obtained through a survey which included responses from 103 oil & gas companies including *"nineteen of the twenty largest US based petroleum companies."* The responding companies had worldwide sales of $583 billion in 1982, which amounted to 92% of the total sales by US oil & gas producers for that year. The survey concentrated on five major areas of interest:

- Investment Measures
- Cost-of-Capital (COC)
- Capital Availability
- Risk Analysis
- Inflation

The analysis of the data regarding investment measures divided the results into discounted and undiscounted methods. Respondents were asked to indicate a primary method and a secondary method. Of the discounted methods, the Internal Rate-of-Return (IRR) approach was favored by 69 companies as a primary method and by 14 companies as a secondary method. The Net Present Value (NPV) method was favored by 32 companies as a primary method and by 39 as a secondary method. The Undiscounted Payout or *Payback* method was preferred by 49 companies as a primary method and by 34 companies as a secondary method. The undiscounted ratio of **Profit-to-Investment (P/I)** method was cited by 18 companies as a primary method and by 30 respondents as a secondary method. The analysis suggests that the IRR, NPV, and Payout methods dominate investment measures.

The Boyle and Schenck study was published by the Society of Petroleum Engineers (SPE) and was widely discussed and cited in oil industry publications. The survey results provide particularly useful information because it concentrated strictly on the oil & gas industry rather than a broader range of industries. The study was also done at a time when evaluation and capital budgeting procedures had moved fully into the computer age, where detailed analysis of both projects and budgetary requirements could be done with speed and variability not available 10 years before. [As an aside: the project evaluation that I presented to Mr. Smith in 1968 was done on a mechanical calculator and a handwritten, 23-column paper spreadsheet.]

In 1993, a similar study was conducted by Dougherty and Sarkar to determine if companies had changed or modified investment evaluation methods and practices since the 1985 Boyle and Schenck study.[61] The 1993 study was also based on a survey format that asked many of the same questions as the earlier survey. A total of 90% of the 107 responses were obtained from corporations, including 6 of the 10 companies with the largest total assets. According to the text of the study report, about 76% of companies used IRR as a primary method and about 24% as a secondary method; approximately 53% used NPV as a primary method and 45% as a secondary method. *"Fewer (92) firms use a non-discounted method in evaluating investments. Of these, 64 indicate this method to be primary and 58 secondary."* The non-discounted method is Payout. Dougherty and Sarkar did not report P/I ratio or profitability index.

The results from both studies suggest that IRR is favored as the primary method followed by NPV as a secondary method, with payout/payback in a supporting role. It is not difficult to come to the conclusion that all three methods may be used to some extent and very likely in combination. Each method provides a somewhat different perspective on a proposed investment, and a procedure that considers the results of all three, or even four methods if P/I Ratio is included, could allow decision makers a more complete picture of potential returns, particularly for longer-term projects. Indeed, it is common to find evaluation software that calculates and presents all four methods as part of a discounted cash flow analysis.

| | Boyle & Schenck | | Dougherty & Sarkar | |
|---|---|---|---|---|
| **Methods** | **Primary** | **Secondary** | **Primary** | **Secondary** |
| Discounted Internal Rate-of-Return | 69 | 14 | 76 | 24 |
| Discounted Net Present Value | 32 | 39 | 53 | 45 |
| Undiscounted Payout/Payback | 49 | 34 | 64 | 58 |
| Profit/Investment Ratio | 18 | 30 | NR | NR |

**Table 5.** Comparison on Boyle & Schenck vs. Doughtery & Sarkar studies

Another source of information regarding investment methodology is the annual Society of Petroleum Evaluation Engineers (SPEE) survey of petroleum industry evaluation professionals. Starting in about 2000, SPEE included data in the survey regarding the evaluation methods used by industry professionals whose primary responsibility is evaluation of oil & gas properties and projects. The 35$^{th}$ Annual Survey (2016) received 190 responses to an extensive questionnaire.[62] Of these, 76% reported DCF (interpreted to be NPV) as the first preference, with only 8% as a secondary choice. ROI or IRR was the primary preference of only 10%, while a total of 68% chose IRR as the second, third, or fourth preference. Only 3% identified payout as a first preference while a total of 59% chose payout as the second, third, or fourth method.

There are some differences in terminology between the SPEE survey and the earlier 1985 and 1993 studies, which may result from either a lack of definition of the terms in the questionnaires or the evolution of common usage in the oil & gas industry. It seems apparent that DCF (SPEE) is the same as NPV. On the other hand, ROI is often construed to be the same as IRR in common usage. The latter is not strictly true, as the terms imply different modes of calculation, but for the purpose of studying investment analysis, the differences can be set aside.

If the SPEE data and interpretation is correct, then NPV would seem to have replaced IRR as the primary method of analysis. In fact, if most commonly used software calculates both NPV and IRR, the primacy of one method over another may be irrelevant. Further, there is no suggestion that any one method is used exclusively. It is reasonable to expect that multiple methods are used in concert. The 2014 SPEE survey noted that, *"Forty-seven (42%) of respondents indicate that they utilized a total of two to five different methods.... The remaining 58 utilized six or more different methods."*[63]

| Preference | Discounted Cash Flow (NPV) | Return on Investment (IRR) | Payout | Profit – Investment Ratio |
|---|---|---|---|---|
| 1st | 144 | 19 | 5 | 7 |
| 2nd | 16 | 62 | 24 | 19 |
| 3rd | 12 | 44 | 39 | 29 |
| 4th | 7 | 23 | 49 | 35 |
| 5th | 1 | 18 | 29 | 36 |
| 6th | 4 | 15 | 20 | 33 |

**Table 6.** Valuation Method Preferences – Number of Responses (Abridged) 2016 SPEE Survey of Parameters

While additional data may be available from other sources, it seems apparent that NPV, IRR, and payout are the dominant methods used for investment analysis by oil & gas companies. This should be expected of survey results taken from knowledgeable companies, evaluation professionals, and investors since these three methods are most prominently considered by persons responsible for financial management and discussed in business management texts.

# Practice and Application

Having defined the methods most used by industry companies to evaluate potential investments, a discussion of the practice and application of the methods is in order. As noted, there are numerous financial texts available for this purpose. The following is borrowed in large part from a standard business/financial text titled *Fundamentals of Corporate Finance* by Brealey, Myers, and Marcus (hereafter BM&M).[64] To wit, "*The investment decision, also known as capital budgeting, is central to the success of the company.... A company's shareholders prefer to be rich rather than poor. Therefore, they want the firm to invest in every project that is worth more than it costs.... Sometimes the firm may be forced to make choices because it does not have enough money to take on every project that it would like*[65] BM&M present their discussion as a set of Rules.

It is important to note a couple of conditions that should be considered when reviewing and using the methods.

1. NPV and IRR are invariably discounted cash flow calculations; they require the use of an appropriate discount rate, usually the opportunity cost-of-capital, to obtain the NPV and as a comparison point for IRR.

2. The methods are not restricted to one cash flow for a specific project; multiple cash flows can and should be run for each project to account for potential differences in production and economic conditions going forward.
3. Payout and (sometimes) P/I Ratio are non-discounted methods, but here again, multiple cash flows would provide greater opportunity to evaluate an investment decision.

## Net Present Value

The (NPV) method compares the present value (at a specific discount rate) of the income stream derived from an investment to the total capital investment required to obtain the income stream.

Important caveat: the income stream must be only the incremental cash flow obtained from the investment and does not include income that would have been received absent the investment.

When calculated prior to deduction of investment, the NPV is compared to the total amount of capital investment. If the NPV exceeds the total investment, the project should be acceptable; if NPV is less than the total investment, the project can be set aside. That is, if a proposed project requires an investment of $1,000,000 and the income stream reduced to present value at X% equals $1,500,000, then the project should be acceptable. But, if the NPV of the income stream is less than $1 million, the project would not be accepted. In the budgeting process, the NPV of each of a series of proposed investments are compared; the projects with the highest NPV are selected until available funds are exhausted. This approach can have variations and anomalies, particularly if other budget criteria are considered or if the company has investment criteria other than strictly financial results; however, the NPV method is a baseline approach.

According to BM&M, the NPV Rule states *"that managers increase shareholders' wealth by accepting all projects that are worth more than they cost. Therefore, they should accept all projects with a positive net present value."*[66]

## Internal Rate-of-Return

The IRR is the discount rate that would be used to:

- reduce the incremental income stream to a present value equal to the amount of the investment, or
- reduce it to zero if the investment is deducted in the income stream.

Many authors have suggested modifications on this method, and variations are in use depending on individual or company preference. In general, the method employs an iterative sequence of discounting the income stream at various rates until a match with the investment is achieved. The method can also be done graphically by calculating the present value of the income stream at various discount rates, plotting the points on a graph, defining a curve through the points, and finding the discount rate at the point on the curve where the present value of the income stream equals the amount of project investment. Most oil & gas evaluation software programs calculate and report one or more forms of the IRR. In application, the firm would accept a project that provided a rate-of-return that exceeded the opportunity COC or MRR. In budgeting, projects with the highest potential IRR are accepted until funds are exhausted.

BM&M define the IRR as, *"the discount rate that gives the project a zero NPV.... also termed the DCF rate of return."* Further, *"Invest in any project offering a rate of return that is higher than the opportunity cost of capital COC."*[67] BM&M note that when NPV and IRR are used in project ranking, the methods should provide similar results.

Typically, the NPV decreases at increasing discount rates. However, some types of oil & gas investment projects (acceleration projects) yield *multiple rates of return.* In the case of in-fill drilling, for example, the total oil recovery and its undiscounted value may not be increased; however, the production is accelerated into a shorter time frame. In these cases, the NPV will be negative at low discount rates, increase to a maximum value, and then decrease with further increases in the discount rate. If the maximum value is positive, the project will have two discount rates that yield an NPV of zero.

## Payout or Payback

The **payout period** for a project is the time (measured as months or years) required to accumulate an amount equal to the investment from the undiscounted income stream derived from the investment. In a financial context, payback is the ROI required by any investor. A common rule of thumb is 3-5 years. It is important to measure payout using only the incremental income stream created by the investment – not income that would otherwise have been obtained. This method was widely used as the primary method of investment analysis until computer-based economic models became generally available. The method is simple to use and has a certain appeal, particularly for relatively small investments and/or smaller companies for whom more sophisticated analysis is not always considered necessary.

The payout method suggests that a project should be accepted if the payout period is less than a specified period defined by the investor. BM&M note, *"The payout rule states that a project should be accepted if its payback period is less than a specified cutoff period."*[68] Application is relatively simple in that all that is required is to total the incremental expected cash flows in the first few years of operation until the total is the same as the amount of the investment. If that balance occurs within the defined payout period, the project can be accepted – if not, the project could be declined.

The Payback method has some serious drawbacks.

1. The method ignores all cash flows after the cut-off date irrespective of the size of the cash flows in later years.
2. The Payout method gives equal weight to all cash flows before the cut-off date even though cash flows that occur in the first year or two usually have higher present value than cash flows in later years. In other words, the Payout/Payback method does not consider the size, shape, and duration of the cash flows that could be obtained from the proposed investment.
3. If the firm uses the same cut-off date regardless of project life, it could lead to the acceptance of possibly poor, short-lived projects and rejection of good, long-lived projects. In budgeting, Payout may be a useful secondary criterion for short-term projects such as well replacements or equipment changes but not for a multi-well drilling program or an enhanced recovery program.

## Profitability Index

The **Profitability Index** is the ratio of the BFIT net present value derived from an investment to the amount of the investment.[69] BM&M define the profitability index as *"the ratio of net present value to initial investment."* Further, *"pick the projects that give the highest net present value per dollar of investment."*[70] Whichever is chosen, it must be used consistently for all evaluations. It was noted earlier that the Boyle and Schenck, the Dougherty and Sarkar, and the SPEE surveys obtain information about oil & gas company investments from industry sources and that these responses are not unique to the industry but are common capital budgeting practices. BM&M provide some further information from a range of industries.[71]

## Profit-to-Investment Ratio

This term is discussed here because it is reported by the Boyle & Schenck and SPEE surveys. The P/I Ratio is used primarily to measure the return on stock investments and is calculated as the ratio of the undiscounted BFIT net income to the amount of investment. The term is often used incorrectly as a synonym for the preferred business/financial term, Profitability Index.

# Investment Decision Making

Every company and individual has a unique approach to making investment decisions. As shown by Boyle and Schenck, by Dougherty and Sarkar, and the SPEE Survey, many companies use two or more methods together. Companies commonly use a particular discount rate, often derived from cost-of-capital as the opportunity COC, which may also be known as the hurdle rate or MRR, to discount all income streams. Assuming some consistency in other income stream parameters, the various projects may then be compared to each other and the most favorable projects selected.

| Investment Criterion | Percentage of Firms that Always or Almost Always Use Criterion | All Firms | Small Firms | Large Firms |
|---|---|---|---|---|
| Internal rate-of-return | 76 | 3.1 | 2.9 | 3.4 |
| Net present value | 75 | 3.1 | 2.8 | 3.4 |
| Payback period | 57 | 2.5 | 2.7 | 2.3 |
| Profitability index | 12 | 0.8 | 0.9 | 0.8 |

**Table 7.** Capital Budgeting Techniques Used in Practice Average Score on 0-4 Scale (0 = Never Use, 4 = Always Use) (The published paper reports 392 total responses)

As an example, assume a company has six projects in which it can invest funds but does not have sufficient funds to go forward with all six opportunities. Evaluations of each of the six projects are prepared by in-house engineers and financial staff based on corporate criteria. Assume that a company has a cost of capital of 16% BFIT and an MRR of 18% BFIT. The company requires that all income streams be discounted at MRR to determine the NPV net of investment and also requires that IRR on each project meet or exceed MRR. In this example, all the projects were assumed to have the same degree of uncertainty. Assume, further, that the company has $70,000,000 available for investment and, further, that the funds are from internal sources; there are no partners and no access to borrowed capital. In this comparison, Project A has the highest NPV and an IRR of 24.29%; the payout is almost 4 years, which may be at the outside of the accepted range but the investment is for five new Marcellus wells spread over eight months. Project F has a considerably lower NPV but a high IRR and a shorter Payout Time. The two projects require total capital of $49,000,000.

| Project | Investment, $ | Total Net Cash Flow, $ | Net Present Value, $ | Internal Rate-of-Return, % | Payout Time, Mos | Profitability Index |
|---|---|---|---|---|---|---|
| A | 40,000,000 | 101,212,203 | 46,964,187 | 24.29 | 46 | 2.53 |
| B | 16,000,000 | 24,769,025 | 16,067,131 | 17.99 | 41 | 1.51 |
| C | 7,500,000 | 13,777,136 | 7,964,412 | 20.76 | 40 | 1.84 |
| D | 5,000,000 | 8,415,120 | 5,410,774 | 22.36 | 41 | 1.68 |
| E | 5,000,000 | 6,273,638 | 5,054,297 | 19.13 | 33 | 1.25 |
| F | 9,000,000 | 11,919,206 | 9,810,271 | 28.23 | 23 | 1.32 |
| TOTAL | 82,500,000 | 166,366,328 | 91,271,072 | | | |

**Table 8.** Six project evaluations used to determine optimal investment

A combination of Projects C, D, and E would have a total required investment of $17,500,000; the total NPV is marginal at $18,429,483, but has a composite IRR that exceeds MMR. A ranking of A, C, D, E, and F would be appropriate. There are circumstances when B could be accepted despite its low(er) IRR, such as when intangible considerations come into play.

# A Note on Common Applications

The methods of analysis discussed above are commonly used in concert. While one company or analyst may favor one method over another, it is also generally recognized that each method has utility in helping to make investment decisions. That is, a company may require that the evaluation of a proposed investment determine the NPV, the IRR, Payout, Profitability Index, and possibly other measures that the company perceives as useful. This approach is readily illustrated in most, if not all, computer software used for economic evaluation. One widely used software package calculates:

- NPV at six different discount rates,
- IRR,
- Payout, and
- Profitability Index

as part of the default standard output format. Most other software packages follow the same approach.

# 16

# REGULATORY COMPLIANCE EVALUATIONS

To this point the evaluation process has been discussed as being used for estimation of future production and income and the present value of that income, for purposes initiated by the operator or other parties with an ownership interest in the property. Evaluations for regulatory compliance follow the same procedures for estimating oil & gas production and income, but have four other conditions in common:

- They require an estimate of the *(Fair) Market Value* of the property.
- They usually relate to some form of tax or transfer to a government agency.
- Evaluation conditions are (often) imposed or at least "suggested" by a regulating agency.
- The evaluation is (often) done in consultation and coordination with an attorney and/or CPA.

## Ad Valorem Tax

Ad valorem is a Latin term which means *"of value."* Taxes levied on the value of real property are one of the oldest means of tax collection. The earliest known tax records were found in ancient Mesopotamia dated to about 6000 years BC. In ancient Egypt taxes were levied against the value of grain, cattle, and beer as well as land. Taxes on property in England were paid in the form of rents, with more productive land having a higher rent. William the Conqueror introduced a form of property tax which was later modified by the Magna Carta; the system was codified later in 1689 to be controlled by Parliament. The English system was carried to America by the earliest colonists and adapted to conditions in the colonies. In the early United States the property tax was, along with tariffs on imported goods, the primary source of revenue for government.

Real property was (and to some extent still is) a source of both political power and of beneficiary services to the owner and was, therefore, the primary subject of taxation. Property was obvious, easily measurable, and, unlike trade goods, not mobile, so the governing authority could establish a value and then set a tax rate equal to a portion of that value. That process has not changed in hundreds of years. Despite a relatively low profile within the petroleum industry, ad valorem tax assessment probably creates more *market value* appraisal work than any other evaluation purpose.

Mineral rights, including oil & gas rights, are recognized as real estate for property tax purposes in most states where oil and/or gas production is economically important. Three states, California, Texas and Utah, assess the ad valorem tax value of mineral rights on individual properties based on periodic appraisals of the *fair market value* of those rights. The assessed value and amount of tax is determined by county assessors (California), appraisal districts (Texas), or the state (Utah). Taxes are assessed against oil & gas properties based on the market value of the future income expected from the property as of a specific date, usually termed the "lien" date.

The form of the tax and the assessment methods differ among jurisdictions. Property tax assessment is a special form of valuation that starts from a market value but then imposes limitations and restrictions. While all three states rely upon market value on a specific date, property tax assessment is governed by state laws, often buttressed by litigation, and is regulated to varying degrees. Texas law requires properties to be assessed at fair market value, but it also imposes a maximum oil price projection or "cap" on assessment evaluation and suggests a range of discount rates. The state, through the comptroller, establishes guidelines and recommendations for use by appraisal districts through the issuance of a manual for oil & gas property assessment and annual supplemental guidance.[72] The state also conducts audits of assessment practices.

California has more specific regulations that, while not imposing detailed assessment parameters, do dictate the form and content of the appraisal. Largely as the result of litigation,[73] the state has enacted regulations for oil & gas assessment practice codified in an extensive manual that defines the preferred sources for derivation of production estimates, product prices, and discount rates.[74] For example: in theory, only the value derived from proved production can be assessed, and the evaluation is to be done before federal and state income tax. The California regulations define the sources of economic criteria such as discount rates but do not specify the product prices or discount rates to be used; the parameters used in annual appraisals are left to the various county tax assessors. This latter point has led to litigation over assessed values wherein the parameters used by assessors have been subject to vigorous debate.

# Eminent Domain

Eminent domain is another ancient concept affecting property ownership. The term apparently derives from a legal treatise *De Jure Belli et Pacis*, written by the Dutch jurist Hugo Grotius in 1625, which stated, in part:

> *The property of subjects is under the eminent domain of the state, so that the state or he who acts for it may use or even alienate and destroy such property, not*

*only in the case of extreme necessity, in which even private persons have a right over the property of others, but for the ends of public utility, to which ends those who founded civil society must be supposed to have intended that private ends should give way. But it is added that when this is done the state is bound to make good the loss to those who lose their property.*

The power of government to take private real property has always existed in the early English colonies and in the United States as an inherent attribute of sovereignty carried over from English law and practices. The power resides in the legislative branch and may not be exercised unless authorized by statute that specifies who may use the power and for what purposes. The legislature may take private property directly; however, it may also delegate the power to government agencies or to private entities such as railroads, public utilities, or even individuals. The basis for the federal government power derives from the Fifth Amendment to the United States Constitution which states, *"nor shall private property be taken for public use, without just compensation."* The Fifth Amendment did not create the federal government's right to use the eminent domain power; it simply limited it to public use. Originally, this portion of the Fifth Amendment, known as the "Takings Clause," applied only to the federal government. The states followed their own interpretations, but numerous court decisions began to invoke the Fourteenth Amendment of 1868 to apply the takings clause to state applications. Over time the application of the takings clause has expanded to allow uses not conceived by the framers of the Constitution including, but not limited to, condemnation of private property by redevelopment agencies for the benefit of another private party under the guise of "public improvement" [see Kelo v. City of New London 545 U.S.469 (2005)].

Eminent domain is inherently a legal process and is far too complex and contentious to be fully discussed here. Suffice it to say that eminent domain becomes an issue when government or a government agency at the federal, state, or local level determines that a private property is needed for some theoretically public purpose. Eminent domain applies to real property, and, since mineral rights are real property, the process can be and has been applied to oil & gas properties. All of the rules and procedures that apply to real estate appraisal apply to any mineral rights that are a part of the taking. For example:

- A proposed street widening in the Signal Hill area of the Long Beach field required the removal of several pumping units on producing wells. The pumping units could not be moved without severing access to the oil being produced. Since retaining access would require new wells, which may or may not be justified by the value of the remaining production, the city also had to compensate the oil & gas rights owner for the lost value of the remaining production.
- The planned development of high-speed rail transportation in the San Joaquin Valley (known affectionately by California taxpayers as "Jerry Brown's Moon Beam Express") included acquisition of a right-of-way

> on property near downtown Fresno. This property and the surrounding area is not now and never has been an oil-producing area. Nonetheless, the California Department of Transportation required an appraisal of the oil & gas rights underlying the right-of-way prior to condemning the property under eminent domain.

When the government cannot reach a value agreement with the property owner eminent domain is invoked and litigation ensues. The US Supreme Court and virtually every state court have heard and decided numerous takings cases that have attempted to define *just compensation* as fair market value or market value, and expanded application of market value in the eminent domain process.

Leaving aside the appropriateness of the application of eminent domain, it is the just compensation part of the process that often requires an economic evaluation of the property being taken. The use of the term just compensation was originally intended to compensate the property owner for the true and just value of the property rather than simply paying the owner some arbitrary amount. Thus, following the condemnation of the property, step one in a taking, the government must offer the owner an amount that compensates for the loss of the property: for this discussion, the oil & gas rights. The owner is not obliged to accept the amount offered and is entitled to submit his own value. The differing values can be negotiated but may end up in litigation where a court must decide the value. Depending on the size of the compensation and difficulty of negotiation, there is opportunity for valuation experts to be engaged by either side.

Even though evaluation for eminent domain is commonly based on the market value of a property, the process has been litigated in federal and state courts to such an extent that it has evolved into a special area of practice for both attorneys and evaluators. It has also developed an entire lexicon of terms and methods of valuation that are not found in other appraisal areas. A large part of appraisal literature, including several entire books, are devoted to estimating the value and amount of appropriate compensation. Further, the procedures for estimating value are often subject to "local" government interpretations of existing law and the application of valuation methodology.

Actual physical takings of oil & gas properties are comparatively rare. The large majority of those cases that do occur are in urban/suburban areas, particularly where urban growth has come into contact with and, as in the Los Angeles Basin, incorporated oil-producing properties resulting in increasing demand for consolidation and/or removal of those properties.

Outside of urban areas, oil properties tend to be impacted more by regulatory takings than by physical takings of a portion of the property. Regulatory takings occur when government actions restrict the operation or beneficial purpose of a property to such an extent that the value of the property is impaired. The impact of regulatory takings is often difficult to measure. Further, the legal character of regulatory takings is unsettled and remains as much a political matter as a legal one, although the distinction between the two becomes increasingly difficult to identify with each new court decision. Nonetheless, a regulatory taking that results in the diminishment or loss of use of all or a portion of a property still amounts to a loss of value and requires compensation. Regulatory takings are a matter of strenuous debate; government agencies that issue regulations commonly do not accept that the regulation causes a loss of use or value, and demonstrating that loss can be time-consuming and difficult.

## Estate and Gift Tax

Estimation of value for the taxes to be paid on oil & gas properties in a decedent's estate also requires estimation of the market value on a particular date. Depending upon the regulatory authority and existing laws, the estimation of value for estate tax would make use of standard evaluation procedures for potential production but may impose certain conditions on the economic analysis, such as limits on pricing projections, operating costs, and/or discount rate. Similar to eminent domain, appraisals of properties for estate and/or gift tax strive to estimate market value in the context of certain restrictions.

Evaluation of market value for **estate taxes** generally follows federal law and regulations[75] that require an estimate of fair market value as of the date of death and/or on the option date six months later. As the result of extensive litigation of mineral values over nearly a century, the methods and procedures to be followed in estimating market value for estate and gift tax are documented in Treasury Regulations such as 1.611-2(d) which deals with oil, gas, and mineral properties. A good discussion of relevant cases and appraisal methods is provided by Posgate.[76]

**Gift taxes** are required to be paid on the market value of property transferred or "gifted" from one party to another (usually related) party, such as from a father to children. The form of the tax and the process of estimating value as of a transfer date are identical to estate tax.

## Transfer Tax

Some states impose a one-time tax on the transfer of ownership of real property including mineral rights. The taxes tend to be relatively minor. The forms and methods for estimating the taxes vary from state to state. They generally take the form of an ad valorem tax but are simplified by not usually requiring a market valuation.

As an example, the Transfer tax in Pennsylvania for Marcellus gas properties is 2% of anticipated revenue for the first few years of production.

# Purchase and Sale Among Related Parties

The purchase and sale of oil & gas properties among related parties, such as transfer of oil & gas rights (royalties) into a family limited partnership or the transfer of oil & gas rights from a non-profit corporation, such as a hunting club, into a separate corporation or Limited Liability Corporation (LLC), is no different than any other evaluation and requires the same care. While not always necessary, a fair market value appraisal is usually done to satisfy all the parties and to provide a third-party opinion of value in the event of state or federal tax concerns.

# 17

# (FAIR) MARKET VALUE AND VALUATION

## Measures of Value – Perspective

The preceding discussion referred to fair market value as it might be applied in the regulatory compliance evaluations. However, before going forward into a discussion of market value and the issues related thereto, it is useful to remember that mineral, oil, and gas properties and rights are generally considered to be real estate. While most real estate evaluation theory and practice relates to surface land, residences, and commercial buildings, many concepts can be carried over to petroleum property evaluation. Real estate evaluation practice recognizes a number of types of value in addition to market value such as Value-in-Use, Investment Value, and Going Concern Value. These are forms of value that may be related to market value and can be applied to the same property. The common difference is that they are values based on a specific set of conditions and not necessarily the value of the property to the market in general.

## Value-in-use

Value-in-use is the value of a property for a specific use based on the productivity of an economic good. "*For example, a manufacturing plant designed around a particular assembly process may have one use value before a major change in assembly technology and another use value afterward. . . . An older factory that is still used by the original firm may have considerable use value to that firm, but only a nominal . . . value for another use.*"[77] It is not difficult to see that oil properties could fit into the value-in-use criteria; oil & gas rights are intended to be exploited and, since the value of petroleum properties is derived from production and sale of oil and/or natural gas, to the exclusion of other uses, an evaluation done specifically for the owner/operator of the mineral rights is a measure of value-in-use to that owner/operator either in its current and intended use or in a potential use subject to investment. Examples of value-in-use evaluations would be internal analysis for:

- property improvements such as for new storage and shipping equipment and/or electrification of lease facilities,
- preparation for third-party financing, and
- reports to WI or other interest owners.

# Investment Value

*"Investment value is the value of an investment to a particular investor based on his or her investment requirements."*[78] Further, it *"reflects the subjective relationship between a particular investor and a given investment."*[79] Translated into the evaluation of oil & gas properties, this is the value that an operator may assign to a proposed property acquisition or the amount the operator might consider contributing to a joint venture for acquisition or development of a property. It is a value that is a function of the investor's particular financial requirements; the project that was proposed to Jim Smith back in 1968 was an investment valuation that did not, as it turned out, meet required investment criteria.

# Going-concern Value

*"Going-concern value is the value of a proven property operation* [and] *... includes the incremental value associated with the business concern, which is distinct from the value of the real estate."*[80] This is an interesting and moderately complex concept. Real estate texts are busy to point out that the value *"includes the incremental value associated with the business concern which is distinct from the value of the tangible real and personal property."*[81] In some respects, the going-concern value can be viewed as an expansion of the value-in-use: the value of the business as a whole rather than the value of an operating factory or oil lease. That may be true in the case of valuing a business. The evaluation of petroleum properties is based on the estimation of the value of the production and sale of oil & gas, which becomes the measure of value of the mineral rights; there is no separate value to the mineral rights absent the production or to the surface real estate. However, the value of a producing property, whether of existing production or production that can reasonably and economically be developed, depends upon the existence and ongoing operations of sometimes extensive surface facilities and access to that surface location. There is no separate value for that equipment or of the business that operates the property; an oil & gas evaluation is a going-concern value.

# Market Value

Evaluations of oil properties done for property sales and/or acquisition, financing, or internal corporate purposes and/or non-investment purposes including, but not limited to, ad valorem taxes, eminent domain, estate and gift taxes, may require an estimate of the **fair market value** of the property. Irrespective of the purpose, an understanding of both the concept of market value and, separately, the application of market value to petroleum property evaluation is necessary and can be considered in two parts:

1. A discussion of the concept and meaning of market value as described by both real estate and legal sources and,
2. An examination of the application of market value to oil & gas property evaluation and some of the methods suggested for estimating the market value of those properties.

# What is Market Value?

> "The central idea in the concept of market value, whether applied to an investment property or a marketable noninvestment property, is the idea of the most probable buy-sell price. The determination of market value is based on the assumption that, if other properties (identical, equivalent, or otherwise comparable to the subject property, as the case may be) have sold at a certain price, or within a certain price range, then the subject property will also command the same price – this is the price a seller will probably get, or the price a buyer will probably have to pay. The validity of the assumption depends upon the continuation of the "market" from which the sales data were obtained, or, more precisely, upon the continuation of the trends demonstrated in that market."[82]

Market value is more a concept than a fact. The concept is that every property has a value in a "*marketplace*" and that if someone wanted to sell a property, there would probably be a buyer who had a reason to buy the property, and, further, that the price that is paid is a measure of the value of the property. The concept assumes that the property can legally be bought and sold. The concept also presumes that the marketplace is amorphous and that any property offered for sale, at any time, would presumably find a buyer. Further, the concept implies that the price paid and accepted establishes a value for that property that may be applicable to other similarly situated properties if those properties were offered for sale at or about the same time under reasonably similar economic conditions.

Market value can be said to become a fact only when a sale of a property has occurred and the price paid is accepted. Even then there may be a question whether the price paid is representative of the market or is conditional to the individual buyer and/or seller. The petroleum property marketplace is a composite of those prospective buyers and sellers who may have the capacity and interest to participate in the acquisition and/or divestiture of oil properties. Put another way, the market is a collection of individuals (persons or corporations), of which only one is actually the buyer and only one is actually a seller. Since the market as such never really buys or sells properties, there can only be individual transactions, which collectively become metaphors for what the market value could be.

If the value defined for a property by an individual transaction is representative of the value that could reasonably be assigned to the property by other buyers/sellers, then the purchase price may become coincident with market value. When the price paid for an individual property diverges from the presumed market value, the purchase price may or may not represent market value. This conceptual conundrum aside, the attempt to estimate the value that the fictional market would place on a property is an evaluation function that is essential for a wide range of purposes. For oil & gas properties, those purposes range from asset sales and acquisition to ad valorem and estate tax requirements.

Market value can be thought of from two viewpoints. For lack of better terms, these are the *Real Estate Appraisal Concept of Market Value* and the *Legal Concept of Market Value*. The two viewpoints necessarily blend together, but it is useful to consider them separately. The term **appraisal** is commonly used to refer to an evaluation that is specifically directed toward estimating the market value of a property. The term is the subject of numerous lengthy texts and learned papers attempting to establish criteria for an appraisal. In this discussion of market value, the terms appraisal and evaluation are used interchangeably.

# Real Estate Appraisal Concept

Mineral, oil, and gas rights are a part of the real estate of a property. There is an extensive body of real estate appraisal literature that is directly applicable to the market value appraisal of oil & gas properties.

As noted in *The Appraisal of Real Estate (ARE)*: "*Market value is inherently a simple concept. It is an objective value created by the collective patterns of the marketplace*."[83] Also, despite differing opinions on individual aspects of the market value definition, it is generally agreed that market value results from the collective value judgement of market participants. This is necessarily a generalization, but it also embodies the basic idea of market value and sets the boundaries for estimating value. It contains two essential ideas: objective value and collective patterns. An **objective value** is one that is conferred by a recognition of the elements of the property that create value without consideration of other issues that may pertain only to a specific buyer or seller. In other words, given a reasonable knowledge of the elements of the property that provide value, what would be the value placed on the property by the market – presumably a number of knowledgeable people acting in individual self-interest? The **collective patterns** refer to a recognition that markets, and the elements of value in a property, change over time as financial conditions vary and as the elements of value change relative to the general economy and the requirements of investors. That is, market value is variable, not fixed, where the variability is a reflection of the collective patterns of economic trends and the requirements of buyers and sellers at a point in time.

This general appraisal concept of market value leads to another basic premise which is: "*For a market to exist, there must be enough buyers, sellers, and product to provide competition; out of this competition a central tendency, a highest tendency and a lowest tendency will be developed.*"[84] That is, in any market there will be some buyers or bidders who would offer a price higher than others, while there will also be those who would offer a lower price. The majority of the market, however, would offer prices or arrive at values that produce a central tendency toward a more or less common value. In the appraisal sense, market value is measured by the central tendency, not the high or low outliers. "*Market value should be the highest possible price represented by the central tendency, it is not...the highest possible price obtainable.*"[85] This means that "*market value results from collective value judgements of market participants, not from isolated judgements.*"[86]

The appraisal concept of market value is the indispensable part of the evaluation process. The concept of property value being set by an objective marketplace through a process that results in a central tendency value is the basis for the appraisal process and is the foundation for the discussion of market value.

## The Legal Concept of Market Value

In a sense, the concept of market value is artificial in that the people who make up the marketplace rarely think of themselves as such – they are simply individual buyers and sellers trying to conclude a transaction. The necessity for a market value concept as a part of property appraisal stems primarily from the desire of third parties, who are often not participants in the marketplace, to be able to estimate the value of properties that may or may not be the subject of a transaction. When Mr. Smith sells a property to Mr. Jones, the value of the property to them is the price they have agreed upon. But as societies have become more complex, people other than Smith and Jones, such as financial institutions and other property owners, along with numerous taxing authorities and government agencies, have developed an interest in that value. Should the transaction become the subject of litigation, the courts and other legal entities become involved.

Fair market value has become a legal concept that has generally been defined by the courts. The result has been the need to formalize and codify a commonly accepted idea, the real estate concept, by attempting to define market value and then endow that definition with a process for achieving an estimate of value that meets the criteria of the definition. Much, if not all, the impetus for creating a structure for market value appraisal has come from the legal and regulatory (government) components. Definitions that are quoted or are contained in appraisal texts tend to derive from court decisions that then become incorporated into regulations and into appraisal language, resulting in the legal concept of market value.

The legal concept takes the generalization of the marketplace embodied in the real estate concept and attempts to describe the conditions that create a market value by using words to create divisions between what is and what is not a valid market condition. The original attempt to define market value is probably very old and more than likely had something to do with taxes.

The Legal Concept of market value has evolved from two continuing societal changes. First, the continued expansion of multiple layers of government and the law into more areas of business from civil litigation to property tax. Second, the tendency of the law to attempt to be ever more specific, which requires increasingly minute definitions. However, the law and the court system being what they are, there is more than one legal definition – every court that has ruled on a valuation case, along with any number of regulating agencies, seems to have felt the need to contribute new prose to posterity.

The result is that appraisers and evaluators must contend with numerous differing definitions of market value. The definitions derived from legal issues do not replace the real estate concept of market value but often have the effect of limiting the manner in which the market value can be estimated. The problem is complicated by a proliferation of definitions for which the language, and the interpretation of the language, change over time amongst jurisdictions and as a function of the intended use. While the real estate concept governs the vast majority of appraisal processes, the two concepts have become interwoven, so that discussion of market value generally begins and, too often, ends with a legal definition

# Definition(s) of Market Value

There are numerous definitions of fair market value or market value (hereinafter "market value"). Many of these definitions exist for a purpose. Professional associations may use a specific definition in order to calibrate the work of their members. Local or state legal usage may require or imply a specific form, particularly in eminent domain or ad valorem tax situations. Despite a growth in the number of words in these definitions, they are not specific; all are subjective, and each may have limited application. For the most part, however, the variations among the definitions are a word here and a few rearranged clauses there. A number of sources, including the United States Supreme Court, have noted *"it is doubtful if the use of the word "fair" adds anything to the words "market value."*[87] The term definition is used here advisedly; it can be argued that the "definitions" are really not so much true definitions as they are descriptions of what a market value should be.

This may be a good place to make two important points. First, the descriptions discussed are not exhaustive; any search of literature for the term fair market value or market value or FMV will return numerous definitions and sources thereof. There is neither room nor energy to include them all here. Second, each of the definitions presented and discussed below were derived in the aftermath of some form of litigation or were adapted from litigation sources to serve a regulatory purpose. Court opinions are naturally backward-looking – they attempt to describe events and conditions related to an event that has transpired, such as a sale of property. Therefore, the definitions of market value are mostly applicable for determining if a transaction that has already occurred was/is market value – according to the prevailing definition. An evaluator can only estimate the market value of a property prior to a sale or acquisition by attempting to emulate the actions of buyers and sellers in the marketplace. He or she can only do so by studying prior transactions and adapting the economic and other conditions of those transactions into an estimate of value under similar conditions.

Market value evaluations for ad valorem tax assessment are often done by a county tax assessor or an appraisal district and are submitted to the property owner along with tax assessment based on the assessor's estimate of value of the property. Evaluations for eminent domain are usually done by the condemning authority as part of the condemnation proceedings. Evaluations for estate and gift tax are commonly done by the property owner and submitted to the Internal Revenue Service and, where appropriate, state tax agencies, subject to review by those agencies. Because differences of opinion as to the value of a property can occur between the property owner and the assessor or IRS or other authority, litigation may ensue.

As a starting point, Black's Law Dictionary (2016) defines "fair market value" as *"the price that a seller is willing to accept and a buyer is willing to pay on the open market and in an arm's length transaction; the point at which supply and demand intersect."*[88]

One legal description that has endured is the Heilbron decision of the Supreme Court of California of 1909. In the decision the Court described fair market value as:

> *"the highest price estimated in terms of money which the land would bring if exposed for sale in the open market, with reasonable time allowed in which to find a purchaser, buying with knowledge of all the uses and purposes to which it was adapted and for which it was capable...the highest sum which the property is worth to persons generally."* [Sacramento Southern R. R Co. v. Heilbron 156 Cal. 408, 104p. 979 (1909)]

The Heilbron decision resulted from an eminent domain case and was based on precedent established by prior cases. The language in Heilbron is succinct yet descriptive, with enough elasticity in the terms and phrases that it does not bind; it is even understandable and, possibly for that reason, remains a standard. In an ad valorem tax case decided by the Supreme Court of California in 1955, the Court said that market value is:

> *"the price that property would bring to its owner if it were offered for sale on an open market under conditions which neither buyer nor seller could take advantage of the exigencies of the other. It is a measure of the desirability*

*translated into money amounts and might be called the market value of property for use in its present condition."* [De Luz Homes, Inc. v. County of San Diego, 45Cal. 2nd 546, 561-62, 1955]

By interjecting an expressed concern for "*exigencies*" and the "*use in its present condition*," De Luz builds on Heilbron. California decisions and citations have been used here because they are familiar, but every other state and local jurisdiction may have similar definitions even if the words are different. Courts in other states have made contributions to the understanding of "market value."

Texas courts have defined market value in several cases. In a 1936 case,[89] market value was defined as:

"*the price the property will bring when offered for sale by one who desires to sell, but is not obliged to sell, and is bought by one who desires to buy, but is under no necessity of buying.*"

Another definition reported by Texas courts stated:

"'*Market value' or 'fair market value' may be defined as the amount of money that a person desiring to sell, but not bound to do so, could, within a reasonable time procure for such property from a person who desires and is able to buy, but is not bound to purchase the property. It is the amount that could be obtained at private sale and not at a forced or auction sale. The terms 'cash market' and 'fair market' value are equivalent to each other.*"[90]

The West Texas Hotel decision also noted that:

"*The real or intrinsic value of property is represented by that sum of money which is a fair equivalent of such property. The proper determination of such real or intrinsic value of property is by the application of sound judgement and common sense in a consideration of the elements which an ordinarily prudent and intelligent business man would consider in the ascertainment thereof.*"

The first part seems to define market value, while the second part lays out the considerations for estimating the value. As noted by Howell,

"*The important principle is to consider everything that would be considered by well-informed buyers and sellers.*"[91]

In City of Austin vs. Cannizzo,[92] the Texas Supreme Court defined market value as:

"*the price which the property would bring when it is offered for sale by one who desires, but is not obliged to sell, and is bought by one who is under no necessity of buying it, taking into consideration all of the uses to which it is reasonably adaptable and for which it either is or in all reasonable probability will become available within the reasonable future.*" [Note the similarity to DeLuz decided in the same time frame.]

The interest of the federal government in market value is primarily related to eminent domain issues brought under Article 5 of the U.S. Constitution, on which the United States Supreme Court has ruled many times. While no single definition appears to be paramount, a "federal definition" of market value has been derived from the language used by the Court in many of those cases and reads as follows:

> *"Market value is the amount in cash, or in terms reasonably equivalent to cash, for which in all probability the property would have sold on the effective date of value, after a reasonable exposure time on the open competitive market, from a willing and reasonably knowledgeable seller to a willing and reasonably knowledgeable buyer, with neither compelled to buy or sell, giving due consideration to all available economic uses of the property."*[93]

Just as state court definitions have varied from court to court and over time, United States Supreme Court decisions have "*been hedged with certain refinements developed over the years in the interest of effectuating the constitutional guarantee*" specified in Article Five.[94]

The U. S. Supreme Court has also decided that while various terms for market value may be employed:

> *"The owner has been said to be entitled to the 'value,' the 'market value,' and the 'fair market value' of what is taken. The term 'fair' hardly adds anything to the phrase 'market value' which denotes what 'it fairly may be believed that a purchaser in fair market conditions would have given,' or more concisely, 'market value fairly determined.'"*[95]

The above relate to federal eminent domain issues. However, other federal market value definitions can be applied in other circumstances, such as estate tax. In Cartwright, the court reported a definition as:

> *"The fair market value is the price at which the property would change hands between a willing buyer and a willing seller, neither being under any compulsion to buy or to sell and both having reasonable knowledge of relevant facts."*[96]

According to one source, the **Canada Revenue Agency** reports the following working definition, apparently derived from comments of Mr. Justice Cattanach in *Henderson Estate, Bank of New York v. M.N.R., (1973) C.T.C 636:*

> *"Fair market value generally means the highest price, expressed in dollars, that a property would bring in an open and unrestricted market between a willing buyer and a willing seller who are both knowledgeable, informed, and prudent, and who are acting independently of each other."*[97]

Another source suggests that the following is generally accepted by Canadian courts:

> *"The highest price available in an open and unrestricted market between informed and prudent parties, acting at arm's length and under no compulsion to act, expressed in terms of cash."* This appears to be an extraction and interpretation from the Cattanach opinion, not the actual words.[98]

# Applied Market Value Definitions

Market value definitions or descriptions enunciated by the courts quickly find their way into appraisal usage, usually by following a path through regulatory language thence into appraisal practice.

> As an example, when Heilbron and De Luz became absorbed into tax regulation, the California Revenue and Taxation Code broadened the language by adding: *"'Fair market value' means the amount of cash or its equivalent that property would bring if exposed for sale in the open market under conditions in which neither buyer nor seller could take advantage of the exigencies of the other, and both the buyer and seller have knowledge of all the uses and purposes to which the property is adapted and for which it is capable of being used, and of the enforceable restrictions upon those uses and purposes."*[99]

Except for the *"enforceable restrictions"* phrase, which would seem to recognize the expansion of regulatory power since Heilbron, the definition is not that different – but one more important condition has been added.

When regulatory definitions are carried even farther, the conditions seem to proliferate, as in the following California property tax definition:

> *"In addition to the meaning ascribed to them in the Revenue and Taxation Code, the words 'full value, 'full cash value,' 'cash value,' 'actual value,' and 'fair market value' mean the price at which a property, if exposed for sale in the open market with a reasonable time for the seller to find a purchaser, would transfer for cash or its equivalent under prevailing market conditions between parties who have knowledge of the uses to which the property may be put, both seeking to maximize their gains and neither being in a position to take advantage of the exigencies of the other."*[100]

The conditions now include not just "enforceable restrictions" but must consider any "exigencies" of the buyer and seller. Carried further, *"in short, market value is the value in exchange under certain stipulated conditions."*[101]

California employs a similar definition for eminent domain:

> *"The fair market value of the <u>property taken</u> is the highest price on the date of valuation that would be agreed to by a seller, being willing to sell but under no particular or urgent necessity for doing so, not obliged to sell, and a buyer, being ready, willing, and able to buy but under no particular necessity for so doing, each dealing with the other with full knowledge of all the uses and purposes for which the property is reasonably adaptable and available."*[102]

The Texas Property Tax Code defines market value as:

> *"the price at which a property would transfer for cash or its equivalent under prevailing market conditions if:*
>
> - *exposed for sale in the open market with a reasonable time for the seller to find a purchaser;*

- *both the seller and the purchaser know of all the uses and purposes to which the property is adapted and for which it is capable of being used and of the enforceable restrictions on its use; and*
- *both the seller and purchaser seek to maximize their gains and neither is in a position to take advantage of the exigencies of the other."*[103]

# Real Estate Appraisal Definitions

The definition of (Fair) Market Value found in *The Appraisal of Real Estate* (ARE) is apparently rooted in Heilbron, but differs in some interesting ways:

*"The most probable price, as of a specified date, in cash, or in terms equivalent to cash, or in other precisely revealed terms, for which the specified property rights should sell after reasonable exposure in a competitive market under all conditions requisite to a fair sale, with buyer and seller each acting prudently, knowledgeably and for self-interest, and assuming that neither is under undue duress."*[104]

The American Society of Appraisers, which certifies appraisers of real estate, business valuation, and mineral properties, has a somewhat different description:

*"The price, expressed in terms of cash equivalents, at which a property would change hands between a hypothetical willing and able buyer and a hypothetical willing and able seller, acting to arm's length in an open and unrestricted market, when neither is under compulsion to buy or sell and when both have reasonable knowledge of the relevant facts."*[105]

In 1990, the federal government required the creation of appraisal standards that came to be known as *The Uniform Standards of Professional Appraisal Practice (USPAP)*. The creation of standards resulted in a description of market value that would be and is now used by agencies that regulate federal financial institutions in the United States. The federal standard, adopted by USPAP, states:

*"The most probable price which a property should bring in a competitive and open market under all conditions requisite to a fair sale, the buyer and seller each acting prudently, and knowledgeably and assuming the price is not affected by undue stimulus. Implicit in this definition is the consummation of a sale as of a specified date and the passing of title from seller to buyer under conditions whereby:*

1. *Buyer and seller are typically motivated;*
2. *Both parties are well informed or well advised, and acting in what they consider their best interests;*
3. *A reasonable time is allowed for exposure in the open market;*
4. *Payment is made in terms of cash in United States dollars or in terms of financial arrangements comparable thereto; and*
5. *The price represents the normal consideration for the property sold unaffected by special or creative financing or sales concessions granted by anyone associated with the sale."*[106]

# Petroleum Industry Definitions

Discussion, let alone definitions, of market value in petroleum industry literature is sparse. Petroleum evaluation texts tend to avoid the subject altogether. Where descriptions are offered, they commonly quote or paraphrase existing real estate or legal sources. Perhaps this reflects the concentration within industry on valuation for internal purposes. Estimation of market value within industry is generally limited to the relatively small segment comprised of consultants, bank engineers, and tax appraisers who practice in specific areas and adopt the definitions applicable to their specialty and jurisdiction.

Jan J. Arps, in writing about oil property valuation and estimation of fair market value states:

> *"Fair market value of an oil- or gas-productive property, as commonly understood, is 'the price at which a property would be sold by a willing seller to a willing buyer, neither being under compulsion to buy or to sell, and both being competent and having reasonable knowledge of the facts.'"*[107]

The Arps quotation is not attributed to a source, but the Arps description is commonly cited in the oil industry.

Another description, offered by Garb, is as follows:

> *"the amount a willing buyer will pay a willing seller, with the property or interest exposed to market for a reasonable period, neither the buyer nor the seller under any compulsion to buy or sell, both being competent and having reasonable knowledge of the facts."*[108]

This quote is also not attributed to any source, but it is similar to Cartwright. It is essentially the same as those offered by Arps and other writers that occur throughout industry literature.

Then there is the always practical Canadian definition, which has echoes of Mr. Justice Cattanach:

> *"A term often used is 'fair market value,' which is defined as the price that a willing buyer would pay to a willing seller if neither is under any compulsion to buy or sell and both are competent and have reasonable knowledge of the facts. Because all of these conditions are seldom met, the term 'market value' rather than 'fair market value,' is more appropriate to describe the value that is set when resource properties are sold."*[109]

These definitions are reasonably consistent and capture the essential components of Heilbron et al. They provide a useful working description for petroleum property evaluation and/or fair market value appraisal. However, the application of either definition, absent a citation or source, necessarily diminishes the weight that might be given to the valuation, particularly if used in a circumstance, such as an appraisal for ad valorem tax, eminent domain, or estate tax, where an authoritative definition is available and may be required in the jurisdiction.

The market value definitions provided here are limited only by the space available for discussion. They are not exhaustive, nor are they universally applied. However, they do provide representative examples of how market value is conceived in the courts, taxing authorities, and in the appraisal profession. Not surprisingly, the descriptions are not particularly different. One question that bears discussion, however, as noted by the Canadians, is whether the conditions and criteria laid out in these differing definitions are absolute requirements to be rigorously met, or are they goals to which a transaction may aspire but not necessarily achieve?

## Variations Among Definitions

There are many definitions or descriptions of market value, but for the most part they are variations on similar language. The descriptions mean that the true value of a property can only be established in the marketplace by the free and informed interaction of buyers and sellers. The actual value of a property can only be known after the property has been sold under market conditions. Even then, the value is usually the highest price paid, not necessarily the view of the market as a whole. The successful acquisition/sale evaluation is the foundation for the other forms of evaluation discussed below. Other non-market considerations aside, an evaluation done for an acquisition should be expected to satisfy the criteria and conditions of the various market value definitions discussed above.

It should not be supposed that definitions such as Heilbron are universally accepted. Heilbron was a California case and, while broadly quoted, has application only to California. As shown above with the Heilbron/DeLuz pairing, definitions can, and do, change with the passage of time as the result of the recognition of market complexity and the need for greater precision in application. DeLuz recognizes issues in valuation that were either not apparent at the time of Heilbron or were not thought necessary to discuss in relation to the case at hand. DeLuz is a modest update of Heilbron and added little to the Heilbron message except for the concern for "exigencies" or conditions, economic or otherwise, that might cause buyer or seller to act in a manner detrimental to self-interest.

Also, as noted above, definitions differ among the states and between state and federal jurisdictions. The definition of market value that is codified in Texas, Pennsylvania or North Dakota may not differ much from the definition in California, Louisiana or Oklahoma, but the differences exist and can be important if the purpose of the evaluation is to estimate market value in one state or in another jurisdiction.

There can also be a difference among definitions brought about by their purpose and use. Market value descriptions used in civil litigation may stem from court decisions in that jurisdiction, but the definition used in other legal contexts, such as ad valorem tax or eminent domain may differ in obvious if not necessarily ways as the result of the codification of assessment practices or condemnation processes.

The difference in definitions discussed above is important to appraisers of oil property market value only if the appraisal is likely to become part of the legal/regulatory/tax process. Given that many of the uses of a market value estimate are directly related to financing, tax, regulatory, or land acquisition issues, it is essential that the correct definition for the intended use, and the governing jurisdiction, be considered and followed. The evaluator should not assume that his or her general familiarity with market valuation definitions will be sufficient; the estimated value must reflect the conditions and criteria imposed by the intended use in the affected jurisdiction. Further, use of "industry" or "organizational" definitions, absent citation, will probably not be acceptable, particularly when litigation is involved. In this context, it is useful to note that USPAP requires the appraiser of a property to state the definition relied upon in the appraisal.

Rather than attempt to compare all of the multitude of definitions/descriptions, a limited comparison of three sources – the Uniform Standards for Federal Land Acquisitions (hereafter "federal"), the Appraisal of Real Estate, 11th Edition (hereafter "ARE"), and Heilbron/DeLuz – will serve as a sufficient starting point.

The functional differences between the federal and Heilbron/DeLuz definitions are minimal, one of quantity of words rather than conditions – with one exception. The federal definition expressly refers to "*the amount in cash...for which in all probability*" rather than "*the highest sum to persons generally*" articulated in Heilbron. On the other hand, ARE discusses five issues in market value definitions:

1. Cash or cash equivalent versus non-cash
2. Real property rights versus real estate
3. Price versus highest price
4. Most probable price versus highest price
5. Equilibrium value versus market price

Four of the five refer, in one way or another, to versions of value or price which, when dissected, really amount to differences in semantics, or whether "highest price" refers to "value" or whether "highest price" is the same as "value." For that reason, only two of the issues are discussed here.

## Price Versus Highest Price

A comparison of the Heilbron/DeLuz, ARE, and federal definitions offers an interesting contrast. The ARE makes an apparently major change from Heilbron by replacing "*highest price*" with "*most probable price*." Note that De Luz simply says "*the price*" without elaboration. The differences among these can be important when it becomes necessary to apply the definitions to actual appraisal work. If one appraiser is operating under the "*most probable*" approach but another appraiser is targeting the "*highest price*," which is the correct value for the property? But Heilbron contains a qualifier which equates "*highest price*" to "*the highest sum...*

*to persons generally*" which goes back to the appraisal concept of value being a central tendency of many values, while "*most probable*" suggests a consensus among knowledgeable people. This would seem to equate the two descriptions.

## Most Probable Price Versus Highest Price

If open market circumstances are assumed where several equally knowledgeable and informed potential buyers evaluate and make cash equivalent offers on a property, can we also assume that the offers will be reasonably close? Possibly, but additional circumstances may come into play. If "*highest*" price is presumed to refer to the commonly accepted economic and market conditions that prevail for properties at the time of the sale, then "*highest*" could be essentially the same as "*most probable*." But, looking again at Heilbron, the opinion contains another phrase, a modifier that states, "*the highest sum which the property is worth to persons generally*." It could be argued that "*highest*" and "*most probable*" are the same when the market as a whole (*persons generally*) rather than a single buyer is considered.

These two subjects are really variations on a theme and have been debated, with dictionary and thesaurus in hand, by generations of real estate appraisers and lawyers. That said, the variations, though subtle, are important depending on the jurisdiction and end-user of an evaluation for market value. The Most Probable vs. Highest Price is an issue that will be revisited below.

# Evaluation for Market Value

The definitions/descriptions of market value and the methodologies discussed above are derived almost entirely from real estate appraisal. Discussions of evaluation of oil & gas properties for market value tend to enlist the various formulations and go forward to estimate a value. This is acceptable as far as it goes, but the result has been a limited but confused concept of oil & gas property market valuation. The following is an attempt to resolve some of the confusion and provide a basis for petroleum property market valuation.

As treated in this volume, **market valuation** refers to the estimation of market value for an individual stand-alone producing property and does not apply to the valuation of properties as assets of or sources of income for companies or corporations where the property may be one of many – and further, may have particular value to the company for a specific purpose. That is, this volume does not address the valuation of a property as an asset of the current operator or owner. In this context, "producing property" refers to a fee estate or leasehold mineral right, the income derived from that mineral right, and all indigenous equipment, fixtures, and facilities necessary to maintain production and obtain that income. A producing property may consist of one or more geographical units, such as leases that are contiguous, adjoining, unitized, or of such similarity that the combination is a logical unit of valuation.

An evaluation for market value is inherently forward-looking. In contrast, the descriptions and definitions of market value discussed above and extracted from real estate appraisal and legal decisions are necessarily looking back at preceding events. The definitions in Heilbron et al. result from litigation stemming from an event that occurred at some time in the past and the court or other body is tasked to decide how to resolve the issue of value from historical circumstances. A market valuation is concerned only with estimating value for a property based on events (production, sales, etc.) that are anticipated to occur subsequent to a designated date.

The evaluation of an oil & gas property for market value is essentially the same as an evaluation for any other purpose with a few important exceptions:

- The evaluation is done for an individual property with a specific description. The evaluation may include more than one property but they must be valued separately unless they are necessarily combined for the purpose.
- The evaluation is supposed to reflect the conditions and circumstances that would be considered by a knowledgeable person in the marketplace, not the perspective of the person or company conducting the evaluation. In theory, market valuation is a special case of investment analysis where the evaluator is attempting to estimate the value not to himself or his company but to a hypothetical group of reasonably sophisticated investors in oil & gas properties. This hypothetical group of investors is termed the marketplace for oil & gas properties, and the value that would be determined and paid by this market is the market value; that is, it is the consensus investment value of the parties that comprise the market. The production projection, product pricing, operating costs, etc., should reflect the conditions that the market would apply to the evaluation.
- The evaluation has a specific purpose, be it eminent domain, ad valorem tax, estate tax or other reason for needing a market value. Market valuations can be done as part of an acquisition program to estimate the values that competitors could place on a property, but care must be taken not to ascribe attitudes to potential competitors that skew the evaluation.
- The evaluation must not only account for the purpose of the evaluation but put it in context of the location of the property and the jurisdiction that will ultimately decide the value of the property for the purpose.
- Except in limited circumstances, the evaluation will often, at some point, involve the contribution of one or more legal experts, which will complicate the process.

# A Review of Methodologies for Estimating Market Value

This is a good point to step back from the definitions and descriptions of market value from the many sources and focus on the methods that can be used to estimate market value. Several authors have provided discussions and analyses of various methods, some in more detail than others. Of those, two sources in particular should be examined closely.

One is a 1989 paper from William E. Strevig, *"Valuation of Oil and Gas Reserves by the Comparative Method"* in which the author gives direct attention to the comparable sales method of valuation.[110] In the earlier discussion of the comparative sales method, as employed in real estate appraisal, it was noted that comparative sales has only limited use in oil & gas property valuation because, unlike residential and commercial real estate that are commonly evaluated using comparative sales, there is no accessible database of information on oil & gas property transactions that would allow the evaluator of a property to make comparisons and adjustments between the available comparative sales and the property being evaluated. As noted by Strevig, *"The comparative approach to value involves determination of a unit value which can be compared to some objective standard. For oil and gas reserves, the generally accepted unit of measure is "'barrels of oil equivalent.'"* This standard, usually expressed as $/BOE, is similar to the standard of dollars per square feet ($/ft. sq.) in residential valuation.

The comparative sales approach deserves some discussion here because, even though most petroleum properties are valued by the income approach using discounted cash flow, many government agencies, such as the IRS, are accustomed to, and tend to rely upon, comparable sales valuations. An evaluator preparing a market valuation that could end up in court or in front of the IRS should at least address the comparative sales method and comment on the use of the method, even if no value is derived from that method.

Strevig goes on to note, *"The comparative approach has several distinct advantages. The reference standard is derived from actual transactions between real world sellers and buyers.... The emphasis on actual transactions is the essence of the generally accepted concept of fair market value."* The Strevig approach relies upon derivation of the comparison standard ($/BOE) from actual transactions and uses as an example a series of transactions that are reported to have occurred in the fourth quarter of 1988 (the sample date is immaterial) which resulted in a median value of $4.95/BOE for (reportedly) proved reserves. Strevig then notes that *"to utilize this factor in any particular situation requires adjustment based on the quality of the reserves compared to the total spectrum of transactions. There are 26 factors that affect prices paid for reserves,"* which he presents in Table 2 of the paper. The list is comprehensive with three main categories: Reserve Characteristics, Owner/Operator Status, and Transaction Structure.

The Strevig approach is the only published serious attempt to apply the real estate valuation method to oil & gas properties and deserves credit for that effort. There are, however, some flaws in the approach, most of which are common to all procedures based on a standard of value and will be discussed below. The primary issue is the reliance on reserves as the defining criteria of value for oil & gas properties. Both the standard of value ($/BOE) and the adjustments assume that the sole purpose of the transactions used as a data source is to obtain reserves; there is no consideration of the income stream to be obtained.

The second source is a 1990 paper by Forest A. Garb titled, *"Which Fair-Market-Value Method Should you Use?"* in which he notes that *"The various methods for determining FMV fall into four general categories: comparative sales, rule-of-thumb, income forecast, and replacement cost."* He notes further that *"all the methods attempt to establish a fair price for a property that protects the buyer against all foreseeable business uncertainties and risks."*[111] The author states that the valuations based on the income forecast are the most sophisticated and accepted.

Garb notes that while the comparative sales approach has support in the IRS and by agencies that are sometimes required to condemn oil & gas properties under eminent domain, the method is not frequently used in the petroleum industry. *"The most significant advantage to the comparable sales method is that it is simple to use. It requires an estimate of reserves, but does not require a time related forecast of how quickly the reserves will be produced"* (there is that reliance on reserves again). He then notes *"The most serious disadvantage... is that it is difficult to account for specific characteristics unique to the property being appraised."* An understatement. There is an obvious inconsistency here that we will come back to later.

Going on, Garb gives considerable attention to four "familiar" rule-of-thumb methods of valuation as follows:

- Price Paid per equivalent barrel of reserves ($/BOE)
- Price Paid per equivalent barrel per day of producing rate ($/BOE/day)
- Profit-to-Investment Ratio
- Current Income Rate for a specified period of time.

Using the first two methods, individually or in concert, requires information on actual transactions and explicitly relies on the premise that transactions are made to acquire reserves. Thus, *"The current price per barrel of oil in the ground must be determined from recent transactions* [where] *... the information is frequently reported in corporate annual reports and in a quarterly report on oil and gas transactions published by information companies."* There is concern that the information obtained is truly reliable and whether methods based on that information provides a sound basis for estimating market value for an individual property.

Skipping over the Profit-to-Investment method, there is a serious quibble whether the current income rate is a rule-of-thumb or an income method. It is commonly used for valuing small interests, such as a royalty, based on actual property income not on sales transactions. [See further discussion below in Evaluation of Royalty Interests.]

Garb gives appreciable discussion to the income forecast method in the same context as presented above but also notes that, *"After the projections of production and revenue are prepared and after all uncertainties are accounted for through the use of adjustment factors... the income forecasts can be analyzed in conjunction with the rule-of-thumb methods."* Here Garb suggests and recommends that the methods not be used individually but be used in concert so that the utility of the methods to provide somewhat different perspectives can combine to indicate a compromise value. This is not an unreasonable approach.

Both the Strevig and the Garb sources, while dated, are valuable for the information and perspective they present and, in the absence of any other methods, can be helpful. However, the Strevig approach and the methods described by Garb, particularly the rule-of-thumb methods, have four distinct flaws that render direct application to market value analysis difficult and, possibly, misleading.

1. The data used to derive a comparison standard (Strevig) or the rules of thumb (Garb) is apparently supposed to be obtained from public reports regarding transactions that have occurred and for which certain data is available. Strevig, in his 1989 paper, specifically mentions published lists of transactions for a period in 1988. Such information used to be included periodically in industry publications such as the *Oil & Gas Journal,* however, the information usually included nothing more than the names of the buyer and seller, the property or properties, and, possibly, the amount of the transaction; there may also have been accompanying articles about the transaction(s) that expanded the information. Other industry publications would also include information about transactions, some more specific that others, but rarely in sufficient detail to allow derivation of much more useful data.

   Additional data could be available to larger consulting firms (Garb), investment banking firms (Strevig), and the sources of financing (banks, third-party investors), which could be used by those entities to develop market indicators, but this would not commonly be public data that could be used by evaluators. In order to make use of the transaction information, it is necessary to make some assumptions about the properties involved, go to a third-party source of production information (state agency records), estimate a volume of future production, ascertain that the amount of the transaction was the full purchase price, and then calculate the comparison standard ($/BOE) and/or similar ratios, if possible. This could be an interesting but time-consuming task that would still leave many questions about the veracity and utility of the derived data. The underlying assumption to this work is that using enough transactions would provide statistical results that subsume any errors in the individual derivations.

2. Is the reliance on "reserves" as the motivating factor in market valuation, whether in the transactions used as source data or in the property being evaluated. The standard of comparison ($/BOE) and one or more of the Garb ratios rely on knowing the volume of oil & gas "reserves" in the transaction, and both Strevig and Garb assume that "reserves" are the reason for the transaction; both refer to the value of "oil in the ground" as a defining circumstance to the transaction. While the acquisition of additional production potential was likely a motivation, particularly in the period when both papers were written, reserves are, by definition, a function of economic production (see prior chapters) so that there must also be an income stream which could be the primary objective; after all, why purchase a property just to obtain oil-in-the-ground if it provides no economic benefit?
3. Is the assumption, implicit or explicit, that the transactions which are the sources of data are market value. The simple fact that two parties entered into a transaction to exchange one or more properties for an agreed upon price in some form and a report of that transaction, in abbreviated form, was published in an industry journal or other accessible source, does not make it a market value transaction; there is never enough published data to reach that conclusion. If the intent of extracting data from published transactions for use in a market valuation of a property for ad valorem tax, estate tax, eminent domain, or estimating value for a possible sale or acquisition, is there sufficient information to determine that the transaction could conform to one or more of the market value descriptions discussed above? Probably not, and, if not, the decision to use the transaction data and the conclusion of value based on that data, could be challenged as unreliable.
4. Is the assumption that the data published about the transactions is sufficient to determine comparability to the property being valued. This is an expansion of the first issue but is specific to the application of the comparable sales (and ratios) approach to valuation. In order to apply the Strevig and/or Garb methods, there must be, at some point, a comparison of the property being valued to the data source properties. The evaluator may have extensive information about the subject property but, as noted above, unless the evaluator also has detailed inside information (personal or second hand/hearsay) about the properties in the transactions, there is no plausible way make the necessary comparison.

This is not to diminish the efforts of Strevig or Garb; their work, as well as that of a number of other authors of similar works, is a useful contribution to evaluation literature. However, for the reasons discussed above, their use in market valuation in 2025 is limited.

The dichotomy of market value is that an appraiser can only estimate the value of a particular property by reference to other properties and/or measures of value derived from the sales of other properties and must, therefore, determine if those other transactions were FMV before using those transactions in estimating the FMV of the property being appraised.

There is another dichotomy that requires exploration and which, once resolved, can provide the means of going forward to develop a method of market valuation for individual oil & gas properties. In summarizing the methods for estimating FMV, Garb notes that, *"Of all the recognized methods, those based on an income forecast are by far the most sophisticated and the most accepted."* Indeed, the first exhibit in his paper (Table 1) is of a discounted cash flow projection for a well in Texas. Accepting that declaration, the dichotomy is that all or a very large percentage of the transactions used as data sources by both Strevig and Garb, and which are cited in many other similar treatises, were based on discounted cash flow projections. Of course, the data provided in most of the transactions is not sufficient to determine the form and/or content of the discounted cash flows underlying the transactions. However, it is a reasonable assumption that one or more income projections were used by both Buyer and Seller in the transaction. That being the case, why bother with any method other than discounted cash flow?

Several authors, including Strevig and Garb, have discussed the utility and the shortcomings of the DCF approach to estimating market value. Because the DCF method is not unique to oil & gas evaluation, many of the critiques of DCF have come from outside the industry, particularly from real estate appraisal. One of the most interesting is from a 1991 paper by Joseph Williams titled, *"An Examination of Fair Market Value Concepts for Oil and Gas Properties"*;[112] the paper was later published as SPE 22620, *"Fair Market Value: Theory versus Practice."*[113] Both papers offer a somewhat different perspective on market value and the methods of estimating value for petroleum properties, along with a critique of market value definitions as they may apply to oil & gas properties in both theory and practice. Given the difficulty of using the comparative sales approach, there is good reason to revisit the discounted cash flow approach and to review some of the issues raised by Williams, Garb, and others.

The Williams analysis was originally written for and was published in *The Appraisal Journal* which, while primarily concerned with surface real estate appraisal, does, on occasion, include articles on appraisal of other forms of property and generic evaluation topics, such as discount rates, that can be applied to oil & gas properties. This is the path taken by Williams who notes, *"Oil and gas properties present unique valuation problems because of the many engineering and production variables associated with each parcel."*[114] Throughout *The Appraisal Journal* paper, Williams works to blend some aspects of real estate appraisal with oil & gas property valuation by noting, after discussing FMV definitions, that:

*"The fundamental assumptions of an efficient real estate market [and a critique as applied to oil and gas properties] are:*

1. *The properties are homogeneous [each parcel is unique];*
2. *There are large numbers of buyers and sellers [there are relatively few purchasers of property at any given time];*
3. *The markets are self-regulating [extensive regulations govern property transactions];*
4. *There is an effective clearinghouse for buyers and sellers [no such clearinghouse exists for oil & gas property]."*[115]

These issues are not new but are stated in a different perspective from previous discussions. Williams discusses the comparable sales method of valuation and offers conclusions similar to those observed by Strevig and Garb, but his discussion of the income capitalization or discounted cash flow method is of most interest. Williams is concerned that, when used as a market value approach, rather than as an investment analysis tool, the DCF will yield a result that is *"subjective in nature"* because of the many variables inherent in the calculation of value and the differences in those variables that would be assigned by each evaluator and the fact that the assumptions will *"probably deviate from actual future events"* and the extent to which the variable assumptions *"differ from those generally employed in the market"*[116] such that the end result of the evaluation will not represent market value.[117] Williams then concludes that DCF should not be used as the sole method of estimating market value. He further refers to the concerns expressed by some authors[118,119] that DCF does not accurately predict actual market values for oil & gas properties.

The critique of the DCF approach by Williams and others recognizes the concerns and issues with the approach that have been discussed in publications and presentations for generations of oil & gas evaluators. It is useful to note that all of the sources above derive from the experience of the 1980s and early 1990s and the discussions, while appropriate as background to market valuation, are also somewhat dated. In the intervening period to today, evaluation methodology, aided in no small part by easy access to computer hardware and software that allows rapid calculation of literally hundreds of unique values using a wide range of variables, has advanced with the strong probability of overcoming the concerns expressed by Williams et al.

On that note, it is time to revisit the Most Probable Price vs. Highest Price conundrum. This discussion, usually referred to as the debate over the "value standard," has generated animated discussion and consumed whole forests in the publication of uncounted papers and treatises in the real estate appraisal business for decades, particularly as court cases and federal/state/local laws proliferate, thereby creating either new definitions or adding to the confusion over existing definitions. There appears to be no end.

A recent paper by Michael V. Sanders, published in *The Appraisal Journal* in 2018, provides an excellent summary of this debate. As noted by Sanders, *"Appraisers are routinely confronted with multiple definitions of the terms 'market value' and 'fair market value,' depending on the purpose of the assignment. Many value definitions in common use are needlessly subjective and in clear conflict with other definitions. Particular problems revolve around the value standard (highest versus most probable) and varied conditions imposed on the hypothetical market, which in many cases, do not comport with realities of the market place."*[120]

Sanders is most useful in reviewing and invoking numerous real estate appraisal sources to illuminate the probable price versus highest price divide and to suggest implicitly if not explicitly that, technicalities aside, it may be a distinction without a difference. Recognizing that the oil & gas evaluator may not have a choice in the language of the definition required for the purpose, it is useful to follow the logic. One cogent observation is that evaluators (appraisers) necessarily deal with a range of values and uncertainty when conducting an appraisal; nowhere is this truer than in oil property evaluation.

Sanders further relates how the understanding and concept of the value standard changed over time by observing that the 1950 and 1962 editions of *Appraisal Terminology and Handbook* incorporated Heilbron almost verbatim. However, the 1975 edition of the successor text, *Real Estate Appraisal Terminology (REAT)*, incorporated a definition of "most probable" sales price as, *"That price at which a property would most probably sell if exposed to the market for a reasonable time, under market conditions prevailing as of the date of the appraisal."*[121] One gets a hint here of a shift from an emphasis on historical events to define market value and toward greater reliance on forward, or at least contemporary, events.

The 1981 edition of REAT retained the same definition of market value but changed the value standard from *"highest price...a property will bring"* to *"most probable price...a property should bring."*[122] Carrying the progression further, Sanders notes that the 9th edition of *The Appraisal of Real Estate* (1987) contains the observation that *'highest price' was originally rooted in the idea that market value should be the highest price represented by the central tendency, rather than the highest price within an entire data range, and that...a definition that includes the word highest is perhaps subject to misinterpretation."*[123] Not the clearest language, but then textbooks are compiled by committees. It is notable that the 11th edition of ARE (1996) (and all subsequent editions) have continued the admonition about market value, noting that *"market value should be the highest possible price represented by the central tendency it was not thought to be the highest possible price obtainable.... Any definition that includes the word 'highest' may be subject to misinterpretation."*[124]

Sanders illustrates the misinterpretation that can occur in the definitions by reference to a Nevada case. In 1993, the Nevada legislature changed the eminent domain law value standard from the "highest price" to the "most probable price" following the prevailing federal definition. However, in a 1996 eminent domain trial in Clark County, the judge awarded value based on the "highest price" standard. Further litigation ensued. The state Supreme Court reversed the lower court and noted that *"highest"* and *"most probable"* are not synonymous, and that *"highest price"* could be cast as speculative and subject to *"misuse and abuse."*[125]

Sanders refers to and quotes several other authors regarding the issue, including some that suggest use of "likely price" or "typical price" rather than "most probable"; there are too many to explore here and most are differences in shading rather than substantive. One final useful reference is from the Uniform Standards for Federal Land Acquisitions (2016) which defines market value as *"the amount in cash… for which in all probability the property would have sold."*[126]

In summation, but without saying so, Sanders suggests that the most probable price… whether *defined statistically as the mean, median, or mode, or an expected range of values – is a fairly objective standard, at least in comparison to highest price, which conceivably might encompass any number above the mid-point of a possible value distribution."*[127] Further, *"Whether expressed as a specific number or a range, most probable price represents the portion of a value distribution that is most likely to occur, based on an analysis of market data. Highest price, on the other hand, is more speculative."*[128]

The progression in real estate appraisal thinking toward the most probable standard, even if not fully accepted in many definitions or in universal practice, suggests – and, more correctly, points – to the path for evaluation of oil & gas properties for market value purposes. The concerns about access or lack thereof to reliable and detailed transaction information renders the use of comparative sales approach, whether as Strevig's direct approach or Garb's ratios, untenable for other than cursory analysis. The DCF approach has some deficiencies as noted by Williams and others, such as the need to account for many variables and the tendency to produce a result that may or may not replicate the market place. However, the logistics of property evaluation, for whatever purpose and in whichever jurisdiction, have been enhanced since the 1980s and 90s through the use of sophisticated evaluation software, whether a spreadsheet program such as Microsoft Excel™ or a customized evaluation program.

There is no question that evaluations can be built that encompass reasonable variations in all the components of an oil property evaluation, including but not limited to: production projections, starting prices, price projections, operating costs, third-party interests, taxes, investment amounts and scheduling, regulatory requirements, and discount rate. The degree to which variations are applied is up to the evaluator. It is possible to envision evaluations that might vary all the components and end up with a large number of value results. Those results when plotted as Price/Value vs. Probability of Occurrence should give rise to a distribution of values that could be Normal or possibly Log-Normal but, in any case, would provide a most probable value or range of values) and a highest value. If the analysis of the components and the application of the variations are done as "market expectations," as others might see the property, rather than from the perspective of the evaluator, then the results could arguably be presented as market value regardless of purpose and jurisdiction.

# A Suggested Model for Estimating Market Value

Going back to some points made earlier, an evaluation for market value is done for an individual property from the perspective of knowledgeable persons in the marketplace for a specific purpose (eminent domain, estate tax) that requires an estimate of market value subject to the location of the property and the jurisdiction which will ultimately decide the value of the property. Irrespective of the intended use, the market value of a specific property can be (and often must be) estimated using an economic evaluation.

Two caveats must be considered. First, while 90-95% of any market valuation is generic and would have universal application, the "properly done" evaluation will be determined by the intended purpose and by the jurisdiction that applies the definition. Second, the intended purpose may require that certain elements of the evaluation be conducted using specific guidelines and/or that appraisal data be obtained from particular sources in a prescribed manner.

With those conditions in mind, a market valuation can be conducted that will provide both a most probable and highest value.

- Define the property including location, production status, and ownership.
- Define the purpose of the evaluation and the controlling jurisdiction.
- Format the evaluation based on a generic market participant, not the evaluator or his client.
- Define the working and royalty ownership interests in the property.
- Construct multiple schedules of existing production using a base projection and rational variations in initial rates, decline rates, etc.
- If additional wells are expected, introduce production from those wells at appropriate times.
- Apply production risk adjustments where appropriate.
- Construct schedules of product prices including initial price and projections for each product (crude oil, natural gas) being projected. The difference is that market valuations for certain purposes may require the use of price/cost projections imposed by third-parties that are intended to approximate those used by the marketplace.
- The NYMEX commodity price trend is often described as market data and is suggested as a reliable foundation for price projections.

- Construct schedules of operating costs along with projections of variations in those costs over the life of the property.
- Incorporate variations in production taxes and any other existing and/or foreseeable deductions from revenue.
- Incorporate reasonable expectations for additional investment for new wells, major maintenance, or other property-related capital expenditures.
- Calculate a series of income stream projections for the property.
- Discount each projection to present value using an appropriate market value discount rate.
- Having done all that, apply the range of variables to each element of the evaluation and run as many income stream calculations as time and budget will allow.
- Plot the results of the evaluations as a frequency distribution. The peak of the distribution should encapsulate the Most Probable Value. Leave any adjustments to management.

The procedure could be further strengthened by assigning a probability range to each point in each set of variables, although it is not clear that such an exercise improves the result.

The overriding valuation guideline is to apply reasonable and rational criteria to each of the components of the evaluation which, depending on the sophistication of the software used, could result in several dozen or even hundreds of calculated values. These values can then be plotted in a Value vs Probability of Occurrence distribution, which can then be used to extract either the most probable value or the highest value, or both, depending on the requirements of the assignment.

The above is not intended to be a cookbook; experienced evaluation professionals should be able to follow the basic idea without necessarily adhering to every step. Two areas of concern are the range of variation in components such as product prices and operating costs and the discount rate.

It was noted earlier in the discussion of an initial oil price that posted prices could vary from day to day or within a month; if the posted prices are used as an indicator of sales price variation, a range of initial prices could be derived. Estimating a range for natural gas prices may require more dexterity.

Observation of the NYMEX commodity price trend for crude oil, for example, shows that while the NYMEX trades contracts for crude oil for dates as far as ten years from the current date, the near-term contract prices trade much more frequently than the contracts that mature in later years. A review of the daily market trends over a period of several days or even weeks before the effective date could provide a useful range of variation in prices. The same would be true of natural gas price trends.

If operating costs are derived using an average of monthly costs over a year or multi-year period, then it should be possible to derive a standard deviation for those costs and use that range to project costs.

The discount rate for market valuations has been the subject of strenuous debate in several venues, particularly those related to ad valorem tax, which has spawned a cottage industry of study and publications. The market value discount rate is, in reality, no different than any other discount rate in that the purpose is to reduce future revenue to a present value. [See Derivation of Discount Rates.]

Market sales information regarding discount rates used in oil property valuation is very limited. Oil property transactions are sparse and, aside from limited information reported in press releases by public companies, are rarely made public. Obtaining marketplace discount rates from such sales is difficult, and the results may not be definitive. There have been a few professional papers that report analyses of sales data and provide some useful guidelines for discount rates. There are also a few studies sponsored by industry groups or published by government agencies that report discount rate data.[129,130] Unlike the marketplace for residential/commercial properties and surface acreage, there is no public database of oil property evaluation data. However, the studies and reports that have been published, while based on somewhat limited data, strongly indicate that the discount rates derived from market transactions consistently exceed the industry cost-of-capital by several percentage points within a reasonable range.

# 18

# EVALUATION OF ROYALTY OWNERSHIP INTERESTS

As explained in the section on ownership interests, royalty interests can come in several forms. The most common is the simple land/mineral rights owner royalty. The landowner royalty is a function of a lease of the oil & gas rights by the owner, the Lessor, to another party, the Lessee. Other forms include overriding and back-in royalty interests. The landowner royalty is typically 12.5-20.0% of gross revenue from the sale of crude oil, natural gas, and/or other products from the property depending on the terms of the mineral rights lease. Royalty interests often require evaluation just as working interest properties. The valuation of a royalty interest in an oil & gas property is done in the same manner and has all the characteristics, requirements, and processes of an evaluation of a working interest. It is usually necessary to conduct an evaluation of the 100% working interest as a predicate to the evaluation of the royalty interest.

*However, there can be certain complications and exceptions that encourage/require additional consideration and which make royalty evaluation a special case of general petroleum property evaluation.*

Discussion of this subject is placed here, after all the foregoing, for three reasons.

1. While the royalty owner (Lessor) and working interest owner/operator (Lessee) have a contractual relationship through the lease agreement and, in theory, share a common concern in the production, sales, and economic viability of the property, there is a difference between the working and royalty interest.

   The working interest owner, even a minority working interest owner, has control of the ongoing operation property, whether drilling of wells, scheduling of production, arranging for sales, and all other aspects of management. The royalty owner has no control over operations and often may not know the intentions of the operator for development and production from the property.

   It is not uncommon for the Lessor/Lessee relationship to be distant – no closer than the delivery of the next royalty check. There are exceptions, particularly where the lease and production operations cover a large property (think West Texas ranch) with a single (or very few) surface and mineral rights owners and where the lease has been drafted in such a way as to require close cooperation among the parties. Examples might include provisions to control use of the surface so as not to interfere with agricultural activities or, as in some Pennsylvania hunting-club leases, to prohibit drilling and other operations during deer season.

However, the lack of control becomes very important for a royalty interest evaluation in two ways. The royalty interest rarely has any information about future development or changes in operations for the property, which makes it very difficult to include value for future wells, etc. Further, the more distant the relationship, the more difficult it can be to obtain timely and useful information, such as operating costs, for the evaluation.

2. The evaluation of a royalty interest is most commonly done to:
   - Estimate the value of the royalty obtained from a specific property for federal and/or state estate, gift, or transfer tax;
   - Establish a value for use in family/estate planning;
   - Estimate a value related to change in ownership of the property;
   - To provide a basis (rarely) for third-party financing.
3. Because royalty evaluations are commonly done for tax and/or legal purposes, such evaluations are usually done to (Fair) Market Value standards prevailing in the jurisdiction and/or prescribed by the relevant government agency, such as IRS. These conditions may influence some components of the evaluation and must be considered.

# Oil & Gas Rights Properties That Can Be Evaluated

Much of the information and valuation suggestions in this section are derived from extensive royalty interest evaluation work over the past fifteen years in Pennsylvania, West Virginia, Texas, and other regions as the result of the rather precipitous leasing and development activity in the Marcellus Shale, Eagle Ford, and similar venues coupled with several decades of royalty valuation of older California properties. Some aspects of the valuation process for royalty interests can seem esoteric, particularly when compared to working interest evaluations, but in most cases the methods used and discussed are driven by necessity.

Royalty interest properties that may be the subject of evaluation can be sorted into four broad categories:

- Producing – In this case the property or a pooled unit that includes the property is producing, there are sales of oil or/gas, and the royalty owner (Lessor) is receiving royalty income. This category is no different, with some exceptions to be discussed below, to the evaluation of the working interest in the same property. At least 90% of all royalty evaluations fall in this category.

- Leased but Undeveloped – The oil & gas rights are covered by a valid lease. Geophysical surveys may have been conducted on the property. There are no wells drilled or scheduled and no drilling pads built. The property may or may not be subject to a pooling agreement. There may or may not be any well-drilling permit requests submitted or approved for the property or nearby properties.
- Leased, Partially Developed/Non-Producing – The term "partially developed" is interpreted broadly. There is no production on the property. There may be production on nearby properties. The property may or may not be included in a pooled unit. There may be drilling permits requested and/or approved, there may be a development plan in place, there may be roads and drilling pads constructed, and there may be one or more wells drilled either on the property, or on a property with which it is pooled. This category requires considerable data acquisition and analysis as well as the use of expertise and judgement.
- Unleased Mineral Rights – There is no lease and no contemplated near-term development. Evaluation of this type of property is rarely needed or justified except in cases where the property is owned in fee (surface and mineral rights) and an estimated value of the mineral rights is to be included with the valuation of the surface real estate. The condemnation of the right-of-way for the previously mentioned high-speed rail project, aka the "Moonbeam Express," required an evaluation of the unleased mineral rights.

The evaluation of the second and third categories of rights has become more common in recent years due to development of oil & gas in Pennsylvania, West Virginia, and other areas where there was little or no previous activity. Mineral rights owners who found leasing agents at their door and who then signed leases began to realize that there was the possibility of becoming the recipients of potentially large amounts of income from both lease payments and, more importantly, from royalties derived from oil & gas production and sales. While generally a welcome prospect, many owners became concerned about possible tax consequences. This concern led to a desire to transfer the mineral rights assets into separate taxable entities such as an LLC, Family Limited Partnership (FLP), and/or a trust of one form or another, preferably before the increased income began. Since most of the avoidance efforts were thought to require a market valuation, methods were developed for evaluation of those assets.

As an example, in four counties in north-central Pennsylvania, there are a large number of private hunting and fishing clubs that acquired logged-off land from timber companies in the 1940s and 50s. As the land became reforested, the clubs built lodges, roads, fishing camps, and hunting areas. Most, if not all, the clubs were organized as Pennsylvania non-profit corporations with shares owned by the club members. The properties are owned in fee simple by the clubs and varied in size from a few hundred to several thousand acres. Many of the clubs became concerned that the potential royalty income from development of the Marcellus Shale and production and sales of gas would jeopardize the non-profit status of the clubs. This concern led to a desire to sever the oil & gas rights and sell them into a separate LLC which would also be owned by the club members. In order to ensure that the sale and transfer of the rights would pass muster with Pennsylvania authorities and the IRS, a market valuation of the oil & gas rights was considered prudent, if not necessary.

Similar concerns led many farming families in the area to set up trusts, family limited partnerships, or other arrangements to insure an orderly control over and dispersal of potential income. These efforts also required a market evaluation of the royalty interest in the property.

## Royalty Interests in Producing Properties

The evaluation of a royalty interest in a producing oil property is no different than the evaluation of a working interest of the same property. As a practical matter, it is usually necessary to do an evaluation of the 100% working interest to serve as a foundation for extracting the value of the royalty interest. Keep in mind that there are three important differences.

1. The royalty interest is not ordinarily subject to operating costs. The royalty interest receives income as a share of revenue from sales of oil and/or gas with no deduction for operating costs. All operating costs are paid by the working interest owner/operator.

   There are two caveats to this statement. In recent years there have been occasions, particularly where large properties are developed for intensive drilling or enhanced recovery, when operators and royalty owners (usually those who are also surface owners) have entered into agreements which augment the terms of the lease. These agreements often allow the operator to charge or pass along to the royalty owner a fee for transportation of produced oil or require the royalty interest to pay a share of fuel and power purchased from third-parties. These conditions usually become deductions from the royalty income by the operator or the crude buyer. Another approach has been to negotiate the removal or reduction of the royalty on produced gas so that the gas can be used for fuel on the property for EOR operations.

   There is also the situation where operators charge royalty owners for "post-production costs" or "marketing costs" even though, in most instances, the lease terms specifically prohibit such deductions. Experience in Pennsylvania and elsewhere has shown that these charges can be with resulting reduction of the royalty income. While some cases have been corrected by litigation, where such charges or deductions are being made, whether thought to be incorrect or in violation of the lease, they must be considered and included in the evaluation of the royalty interest.

2. The royalty interest does not pay any of the capital investment needed for development and maintenance of the property.
3. The lack of control issue noted above must be considered and accounted for where necessary.

## Assembling Data

Probably the area where the process of evaluation of royalty interests departs most materially from that of working interests is access to and assembly of data. Use of the Income Approach is dependent on the availability of information on production and sales of oil & gas, operating costs, scheduling of new development or, conversely, termination of production, and any potential changes in operations that could potentially affect royalty income. With the exception of very large properties where the royalty owner is also the surface owner and has a cooperative relationship with the working interest (operator), it is often difficult to obtain sufficient data from the operator, particularly operating costs, for a royalty evaluation.

In the evaluation of a royalty interest, the most important information source is the detail statement(s) that accompanies the monthly royalty payments. Depending on the format of the statement, information such as well names, sales volumes, sales prices, ownership interests, and any deductions can be identified, extracted, and used for an evaluation. If royalty payments are being received from more than one company, the formats may be different but the information is usually the same. However, the statement data is often limited to sales volumes (which may or may not equate to production volumes) and does not usually provide performance information such as number of wells, days of production, or production of water and/or natural gas (unless the latter is sold).

Accumulation of data from the monthly royalty statements over two to three or more years can provide a reasonable foundation for an evaluation. It may be the only data available. Many royalty owners track their income and the performance of the properties in which they own interests in a Microsoft Excel™ or other spreadsheet. Even so, a review of the actual statements may reveal information not recognized by the royalty owner.

When possible, the royalty statement data can (and should) be augmented by obtaining additional production and operating information from public and/or commercial data bases. The key to efficient use of a public database, such as the Texas Railroad Commission (RRC) website database, or a commercial site, is to obtain the American Petroleum Institute (API) number for each well from the operator of the property. The API number is unique to each well and is easier to work with than the well name. Depending on the size of the property and number of wells, most companies will provide this information to royalty owners; some require more persistence than others. Experience demonstrates that the operator is often more willing to provide API numbers rather than part with actual production/operating data.

Data on many public agency or commercial websites can be obtained by a search using nothing more than the API number. Depending on the quality and ease of access of the website, the performance history and a wide range of other information for each well can be accessed, downloaded, and used either individually or compiled into well groups. The data will usually include production volumes and days of production. When API numbers are used, the data is provided by well. Keep in mind that the individual well data may be the result of allocation by the operator or purchaser so that aggregation may be necessary.

Further, many of the public websites such as those for Pennsylvania or California include access to well records such as drilling and workover history, not just production information. While the quality and quantity of data along with ease of access varies greatly, the increasing interest of the state and federal governments in oil & gas operations has fostered a desire to collect and make more data available.

## Supplemental Information

The royalty statements provide substantial information; however, it may also be prudent to obtain copies of the original lease agreement along with any amendments and/or revisions, Division Orders, pooling declarations, notices of plans to drill wells, and any unitization agreements that include the property. These documents may, and probably do, contain information that could have an effect on one or more aspects of the evaluation. For instance:

- The oil price reported on the royalty statements for a California property was thought to be correct, but a reading of the lease found that the price to be used for royalty payments was to be the higher of an average of the monthly posted prices for four major crude buyers for similar crude oil or the actual oil sales price, before any adjustments.
- The royalty being paid on a group of Pennsylvania wells was based on the original pooling declaration, but reading of a second and third amended declaration found that the Division Order covering the wells had not been updated for an expansion of the Unit and an increased share for the subject property.
- An IRS estate tax assessment of a Pennsylvania royalty interest included production and sales for six wells for which the operator filed, but then rescinded, drilling permit applications. The royalty owner had no control over the plan to drill or not drill the wells.

It should be noted that these situations could just as easily occur in a working interest evaluation so that review of all pertinent documents is always important.

# Economic Evaluation Methods

There are two useful approaches to valuing royalty interests: the Discounted Cash Flow (DCF) and the Average Monthly Income (AMI).

## Discounted Cash Flow

The Discounted Cash Flow (DCF) works just as well for royalty interests as it does for working interests subject to data availability. While the estimate of future oil and/or gas production is the essential part of any property evaluation, it is the projection of future product prices, operating costs, deductions from royalty, and taxes that converts the production into income and value. Evaluate for the 100% working interest to estimate economic limit, then derive the royalty interest value.

### Product Prices

Oil & gas prices are projected from an initial price, usually the average price in the months prior to the evaluation date. The prices can come from the royalty statement or from another source such as posted prices or information from the purchaser. When prices have been particularly volatile, an average over several months might be prudent along with a comparison to historical trends. Review lease agreements.

For a market valuation, the use of a market-based projection of prices is usually acceptable; the NYMEX commodity futures market is a readily accessible source of market data. Projection of the initial product price using the NYMEX trend for oil and/or gas prices at or about the evaluation date is a reasonable approach.

### State and Local Taxes

The royalty interest may be subject to the same severance, production, and ad valorem taxes as the working interest. It is common for these taxes to be paid by the working interest and deducted from the royalty revenue distribution; however, there can be exceptions.

### Operating Costs

It was stated above that royalty interests are not subject to operating costs. This is true; the royalty interest does not pay operating costs. However, the future income potential to the royalty interest is a function of the continued economic operation of the property. It is therefore important that the royalty evaluation be able to estimate the economic limit, which in turn requires some indication of operating costs and the potential end-of-life of the property. As noted, operators are generally reluctant to provide this data. In lieu of calculating an economic limit, running an evaluation to a specified life of 12 to15 years can be acceptable.[131]

## Deductions from Royalty

While a royalty interest may not be charged for the costs of operation of a well or property, there may be certain deductions from the royalty income that are either allowed by the terms of the lease or are imposed by the operator. Deductions from royalty are more common for natural gas production than for oil and include charges for gathering, treating, transporting, marketing, etc. In many instances, particularly in the Marcellus, some operators have been aggressive in their interpretation of lease terms and have found ways to circumvent lease restrictions. It is not uncommon for charges of several dollars per MCF of gas to occur; there have been periods of low gas prices when the deductions exceeded the gas price. These deductions must be considered in evaluating the royalty interest.

## Present Value

Unless the evaluation is for a specific purpose requiring a market value discount rate, a royalty evaluation can be based on the same present value factor that would be used for a working interest evaluation. There is no validity to the argument that royalty interests carry lower risk and should use a lower discount rate. If anything, because of the lack of control over property operations, the revenue stream to the Lessor has a greater risk and could arguably have a higher discount rate.

# Average Monthly Income Approach

The Average Monthly Income (AMI) approach is simple and, for that reason, useful. It is an offshoot of the payout/payback method of valuation. Assuming that the data is available, the usual procedure is to calculate the average net royalty income (after all deductions) over a 36- to 48-month period prior to the effective date, then multiply the average by 36 months and 60 months, and to use the average of the two calculations.[132] There are some obvious flaws in this approach, but it is simple and has certain applications.

The AMI is easily applicable to evaluation of properties with relatively small income; as a practical matter, a royalty interest receiving less than $3,000 annually does not justify a full evaluation. The IRS will accept the AMI approach for most purposes if supported by income records.

The AMI can be used as a check on the DCF valuation, particularly when there is no operating cost data and the economic limit is estimated by restricting the remaining life of production. If the AMI result is reasonably consistent with the value obtained from a DCF, then both values may be acceptable.

# Evaluation of Leased, Non-Developed/Non-Producing Properties

Properties that are leased but have no development or production are not commonly appraised. However, special cases can arise that require development of a method of evaluation.

> As an example, the extensive and well-publicized leasing of oil & gas rights in Pennsylvania in the 2007-2013 period led to a desire to value these rights to establish a tax basis for newly formed entities (LLCs and FLPs) and for transfer and gift taxes. Since there was no production or income from production, there was no basis for imposing a value based on speculative future production and income.

However, it was recognized that the rights had acquired a value due to the leasing of those rights and it also became apparent that some imagination was required to estimate a value for the oil & gas rights.

The methodology that resulted from the collaboration of local attorneys, accountants, and appraisers and which subsequently became accepted is to value the oil & gas rights according to the terms of the lease. Depending on the date of the lease (lease terms became more favorable for the Lessor over the period from 2006 through 2010) and the negotiating ability of the Lessor (and his attorney), the lease could provide income in the form of a signing bonus, delay rentals and, when and where applicable, an extension payment. The negotiation of the terms of the (lease) contract imparts an enhancement of the value for the Lessee (for access to the property) and value for the Lessor that justifies allowing that access; the Lessor also retains the potential for future royalty income if development occurs.

Further, the value of the subject rights is solely a function of the existing lease terms and is not influenced by the bonus and rentals paid for leases on other properties. The mineral lease is a contract between the Lessor and the Lessee for a specific purpose and duration. The terms of the lease are not (commonly) subject to re-negotiation during the life of the lease. This approach has been used to value hundreds of properties in Pennsylvania since 2008. It has been subject to review by attorneys and tax professionals and has not been challenged by IRS or Pennsylvania authorities. The method is euphemistically known as the "lottery ticket or raffle ticket approach." The Lessee and the Lessor have negotiated and agreed upon terms that set the value of the *raffle ticket*. For the Lessor, the lease agreement represents the potential for a substantial payoff in royalty income, with likely better odds than the lottery; for the Lessee the lease agreement provides the option of development and production with potential future income.

As an example: The oil & gas rights to a property are encumbered by a lease signed in 2008 with a primary term of five years along with a bonus of $100,000 and annual delay rental equivalent to $2,000. The rights could be valued in 2008 at $100,000 plus the present value of the expected delay rental payments over five years ($7,950 at 10%), resulting in a total of about $107,950. Until production and royalties begin, this is the only source of income and value from the rights.

# Evaluation of Leased and Partially Developed/Non-Producing Properties

It is not uncommon to be asked to estimate a value for a property that is "partially developed" but may not be producing. In almost all respects, this circumstance, from the royalty owner standpoints, is not much different from the Leased, Non-Developed situation discussed above. The difference is that some events may have occurred that bring the potential for development, production, and income closer. The extent of "partial development" of the subject property must be defined.

- Are there well or wells that have been drilled but have not been placed in production?
- Are there any wells drilled on adjacent properties?
- Are there producing wells on adjoining properties?
- Have drilling permits been issued for the subject property and/or for adjacent properties?
- Has the operator issued a Notice of Development for the property?
- Are there well pads built on the property?
- How far away is the nearest producing well?
- Is it operated by the same company as the subject property?
- How far away is the nearest oil or gas gathering pipeline?
- If there are no wells drilled or in progress on the property, how much time remains in the primary term of the lease?
- Is the property part of a "pooled unit" and, if so, what share of the production and sales from the unit would be paid to the owner of the oil & gas rights?

The answers to these and any other appropriate questions could have a bearing on whether to attempt to impute a potential income stream to the property. If there are wells drilled and producing on adjacent properties, can that production be reasonably translated onto the subject property even though there may be no wells drilled or scheduled or permitted? There are numerous combinations of circumstances that could justify the creation of a potential income stream subject to the application of a lot of judgement and with the caveat that the royalty owner has no control over the development process, so that from the Lessor's standpoint, any estimate of production and income is speculation.

It may be tempting to say that since the Lessee may have made some efforts toward development, such as building a well pad or running a seismic survey, there is value added to the property. But that value, if any, does not accrue to the royalty interest. Further, depending on the language of the lease, such actions could be construed as operations and could take the lease out of the primary term of the lease. This could result in the cessation of delay rental payments and/or remove the possibility of any anticipated bonuses from extension of the lease to the Lessee or re-leasing of the property to another Lessee. As a practical matter, the degree and form of partial development has little or no influence on the value of the oil & gas rights. Absent a change in the lease terms or the start of production, there has been no change in the income to the rights owner (Lessor). Unless there are exceptional circumstances, in the absence of royalty income, the value of the oil & gas rights reverts to the terms of the original lease agreement along with any amendments or revisions. The only income to be received is the lease signing bonus and delay rental payments. See the example above.

However, assuming the lease is still in the primary term, certain other events might occur that could affect the value of the rights.

- *The Lessee does not drill but negotiates to renew and/or extend the lease for another period of time.* When there is no provision in the lease for automatic extension, the Lessee would have to negotiate a renewal or extension of the existing lease or a new lease. The terms of a new/renewed/extended lease could include a bonus payment, increased royalty, or increased rental payment. The amounts of these payments cannot be estimated with any certainty. The estimation of the terms and duration of the new lease are subject to speculation, with too many opportunities for material error, and would not likely provide a reliable estimate of value.
- *The Lessee does not drill and allows the lease to terminate.* A valuation under this circumstance anticipates that, on the expiry of the primary term, the Lessee does not extend or renew the lease but allows the lease to lapse. At that time, a new lease could be written with another Lessee that might bring a bonus payment and/or additional delay rental payments at a market rate, thereby creating a new income stream that could be valued.

  Given the volatility of observed leasing offers and the consolidation of leased-land positions by operators, the estimation of the terms that might be offered at some point in the future cannot be done with any degree of accuracy and, as with the valuation of future royalty revenue, would include too many opportunities for material error, and would not likely provide a reliable estimate of value.

- *The Lessee finds uneconomic production and/or determines that a well is a dry hole and terminates further drilling.* In this case there is no royalty income to be valued and a diminished likelihood of re-leasing the property to a third party. The Lessor's interest, therefore, has some value or no value, but neither circumstance can be known as of the evaluation date nor can the value of any future income, if any, be determined.

# On the Use and Misuse of So-Called "Sales" Data

There has been some pushback on the lottery ticket approach by some evaluators who argue that the value of the non-developed lease can be determined by comparison to the value derived from the sale of other similar, leased properties or an estimate of the value based on the possible terms of a new, replacement lease. The latter is a non-starter; the existing lease is a contract that cannot be replaced by another set of terms unless the lease is terminated by the Lessee, at which time the property would be available for a new lease.

Derivation of a value for the subject property based on the timely sales of similar leased properties is an attempt to use the comparable sales approach to value non-developed oil & gas rights. This method is common in the evaluation of residential and farm properties as well as business and industrial properties. However, the method fails when applied to oil & gas properties due to the lack of the necessary data to conduct a proper comparison between the lease terms governing the comparable properties and those governing the property being appraised. The most commonly offered sources of information for this effort are the legal filings in the resident County which consist of only basic information.

In order to properly use the comparable approach, each transaction considered useful as a comparable property transaction must meet certain basic criteria:

- The Property must be defined and described as to location, size, etc.
- The Actual Purchase Price for the Property (preferably, the oil & gas rights) must be known.
- The status of the Oil & Gas Rights (leased, un-leased, pooled, etc.) must be known.
- If the property is leased, a copy of the complete lease (not the Memorandum of Lease) including all addenda, exhibits, and attachments is necessary.
- There must be sufficient information on each comparable transaction to allow the properties to be compared to each other and to the subject property.

- No lunch counter data (otherwise known as "hearsay" or "gossip") is acceptable.
- Adjustments must be made to the "value" of each of the comparable properties as compared to the subject property.

The following is a partial list of the characteristics that must be identified for each comparable sale.

- Location – Township, County
- Acreage of Comparable Property
- Original Lessee
- Current Operator, if Different
- Date of Original Lease and Remaining Term
- Date of any Extension(s) and Remaining Term
- Bonus Amount Paid
- Delay Rental Amount Paid and Remaining Amount Due
- Extension Payment Amount
- Royalty Rate
- Standard Lease Form or Modified by Additions
- Purchase Price for Property
- Did the Purchase Price Include the Surface Real Estate?
- Value of Surface Encumbered by the Oil & Gas Lease
- Payments for Surface Use and/or Damage
- Subject to Deductions from Royalty for Gathering, etc.?
- Is the Property Pooled or Unitized?
- Division Orders Issued?
- Well Drilling Permits Requested? Approved?
- Is the Buyer Knowledgeable?
- Was the Seller Knowledgeable?
- Any Restrictions on Accessibility to the Property?
- Any Nearby Production?

Each of these characteristics must be compared to the same characteristics of the property being valued. Where there are differences, the value of the comparable property must be adjusted to account for each difference. If there are five comparable properties, the value of each must be adjusted. The five sales can then be used to derive an indicator of value (such as $/acre) which can then be applied to the property being appraised to estimate a value.

> As an example the subject property has a royalty of 12.5% while a comparable property has a 16% royalty? How much of the purchase price of the comparable is due to the difference in royalty rate? What if the other comparable properties all have different royalty rates?

Experience shows that, except in very rare circumstances, these criteria cannot be met; the method is not valid for valuation of oil & gas rights. The only valid method is the income approach.

## Evaluation of Unleased Mineral Rights

The evaluation of oil & gas rights that have not been leased and may not have even been tested or defined is a rare event. Common practice is to assign a nominal value per acre or some other generic value. The standard form oil lease often contains the term "for value received" of $10 or something similar. Absent an extensive geologic/geophysical assessment, any other value would require a degree of speculation that is probably not appropriate.

# 19

# SIX EASY PIECES

The title and the idea for discussing a set of fundamental evaluation principles was borrowed from a book titled *Six Easy Pieces*, published in 1994, which transcribed a series of lectures by Professor Richard P. Feynman (1918-1988).[133]

## Geology

It may seem elemental and, indeed, unnecessary to say, but petroleum evaluation from drilling to production analysis through economic evaluation of a property is dependent upon and inseparable from the geology of the property being evaluated, that is, its geology in the broad sense of the sedimentation and stratigraphy of the reservoir and source rocks, the folding and faulting (structure) that allowed the petroleum accumulation to occur and persist, and the relation of the reservoir rocks to the geology of the nearby region. With few exceptions, petroleum deposits are found in sedimentary rocks, the composition of which has a direct influence on the potential production and, thereby, the income stream that creates value.

The geology of the subject property includes:

- The sedimentary composition (lithology) of the reservoir formation (sandstone, shale, siltstone, limestone) and the stratigraphy of the formation, the degree of mixing or interbedding of the various sedimentary components. Is the formation a thick section of sandstone or is it interbedded with siltstone or even shale intervals? Is the limestone naturally fractured or of a vugular variety? Are the interbeds likely to be impermeable to fluid flow?
- The structure (folding, faulting, displacement) of the reservoir formation. The ideal is a comparatively flat (level) formation of porous and permeable sandstone with maybe a few interbeds of impermeable shale to help control fluid flow. The ideal is rare. Folding, as the result of tectonic activity, imparts a dip angle to the reservoir that can, and always does, influence the fluid flow dynamics. Faulting, often found to accompany folding, can create obstructions to fluid flow and require development of multiple "fault blocks" in the reservoir (see Wilmington field – Long Beach Unit). Faulting and folding can result in displacement

of portions of the reservoir and inhibit efficient development of the property.

- The nature of deposition (deep water, near-shore, continental, aeolian, turbidity, deltaic) could, and often does, influence the manner in which the property can be developed and exploited, as well as the investment required for development, and the costs of operation. For example, a reservoir made up of lenticular unconsolidated sandstone/siltstone may be a good candidate for cyclic steaming, but, because of variance in thickness and sand/siltstone distribution, may not be an efficient objective for continuous injection.
- The fluids content of the reservoir (oil, gas, water), the comparative distribution of the fluids, and the relation of the reservoir to the nearby formations that might allow access to external sources of energy such as water drive.
- The depth and other potential conditions that will determine whether and how the property can be developed.

Without the knowledge necessary to understand and interpret the geology of the property and its regional relation to other properties, the engineering and evaluation of the property is incomplete and out of context.

## Evaluation of Income Streams (The Jim Smith Rule)

The value of an oil- and gas-producing property lies solely in the present value of the income that can be derived from that property up to the point where the property is no longer earns a profit. Economic evaluation of a property attempts to define the conditions related to the production of the property, such as production rate(s), product prices, costs of production, required investment, and other financial/economic factors that, when combined in a rational analysis, provide a basis for investment decision making. Petroleum in the ground has no value unless it can be produced and sold. Evaluations done to estimate future production and ultimate recovery are speculative unless and until a rational economic analysis can show that the property can be produced with a reasonable likelihood of success and that the production from the property will be sold and obtain sufficient revenue to cover all costs of production and provide a return of and on investment. Such an analysis is necessarily an estimate of value of the future income stream, not the production itself or an estimate of oil and/or gas in the ground. Absent a positive income stream and aside from any potential real estate usage, the property itself has no value and, in fact, could be a liability unless appropriate provisions have been made for efficient retirement of the property. The Jim Smith Rule defines this approach simply: the development and operation of petroleum properties is strictly for the purpose of producing a positive stream of income which has value today.

## Radial Flow and Hyperbolic Decline

At several points in the text, it was noted that this was not a petroleum/production engineering treatise. That said, it must be recognized that reservoir and production engineering concepts directly influence, if not control, the economic potential of a property or project. Of these, the radial flow concept, and its offspring, the hyperbolic decline model, are most important.

The **radial flow equation (RFE)** is one of the foundations of petroleum reservoir and production engineering. The radial flow equation is a useful model for fluid flow through porous media in petroleum reservoirs. The equation is a model making use of a set of ideal circumstances based on laboratory and computer analysis. That model is subject to modification to accommodate observed geologic and reservoir conditions on which the model is based, resulting in numerous variations in the form of the equation.

$$q = \frac{7.082kh\left(p_e - p_w\right)}{\mu B_o \ln\left(r_e / r_w\right)}$$

Irrespective of the variation applied, the model attempts to account for some of the more prominent reservoir properties – permeability (k) and thickness (h), the properties of the principal fluid (oil), viscosity ($\mu$) and formation volume factor ($B_o$), the relative pressure differential between the reservoir pressure ($p_e$) and the wellbore pressure ($p_w$), and the estimated ratio between the "drainage radius"($r_e$) and the wellbore radius ($r_w$) – with the objective of estimating a production rate (q). Of course, as noted by some of the many variations, the model represents an ideal circumstance that does not exist in nature. Reservoirs are not uniform; fluid flow is a function of variations in porosity, permeability, rock composition, deposition geometry, sedimentary boundaries, gravity in thick sections, etc. While pressures ($p_e$) should be reasonably consistent throughout the reservoir, drainage boundaries ($r_e$) are subject to both geologic and mechanical (offset wells) intrusions.

The application of the RFE to economic evaluation has two parts: the estimation of primary production and the estimation of the effect on the production rate; the size, shape, and duration of the income stream; and the value of the income stream brought about by changes in operating conditions, such as:

- Reduction in bottom-hole pressure through addition/improvement of pumping equipment;
- Reducing or increasing the completion interval in wells;
- Employing methods (diluent, cyclic steaming) to reduce oil viscosity in the near wellbore region;

- Enhancing permeability through fracturing or other methods;
- Augmenting reservoir pressure ($p_e$) by means of fluid injection (water and/or steamflooding).

Fluid flow in porous media and the size, shape, and duration of production of oil and/or natural gas are governed by physics and, to a degree, chemistry, which can be influenced but not controlled by the operation of the property. As should be expected from the RFE, virtually all primary production follows a hyperbolic form of decline:

$$q = q_i \left(1 + D_i t\right)^{\frac{-1}{n}}$$

As noted earlier, other forms of production decline, exponential and harmonic, are special cases of hyperbolic decline. The importance of an understanding of hyperbolic decline has been demonstrated in the development and production, and subsequent economic evaluation, of properties in the Marcellus Shale and other fractured shale fields. Production of oil or gas from these properties demonstrates sharp hyperbolic declines in the first 12-36 months of operation before beginning to flatten in later years.

# Things Change

There is a country-western song from the 1980s with the refrain *"everything changes, everything changes,"* which could well be a metaphor for oil property evaluation. Investments in petroleum properties and projects are long-term ventures often requiring years to recover the initial investment and decades to attain the full potential of production and income. Evaluation of oil & gas properties requires projections extending over 10-20 years or more, not simple quarterly or annual estimates. During such a long period, the physical conditions of fluid flow in the reservoir and the production performance may change modestly, but the economic aspects, such as product price, operating costs, production taxes, and regulatory interference, can and do change frequently and often to a significant degree. Product prices are the most notable economic function subject to change, but the costs of production and other economic conditions can, and very likely will, also change. The potential for changes in the economic conditions must be recognized and considered in the evaluation of properties.

- Product Prices are defined, for the most part, by supply and demand for crude oil and natural gas in world and regional markets with the caveat that the market is often disrupted by external interference by governments and/or other interested parties. Supply and demand may be somewhat definable and predictable, but there is always the unexpected influence of natural disruptions, abnormal weather, pipeline disasters, etc., that may be short-term and manageable. External interference is another matter. The answer to price projection uncertainty is that none of the potential disruptions are new or unique; historical price performance can be accessed to model future prices. The commodities markets provide a rational basis for product price projections.

- Operating Costs tend to be more stable than product prices and can often be projected based on reasonable expectations for long-term industrial inflation (producer prices not consumer prices). Where there is sufficient differentiation of operating costs (fixed, variable, fuel, etc.), future changes in expected costs can be projected using different but relevant criteria.
- Production Taxes are a reflection of government regulation and should be assumed to always increase at rates derived from historical precedent.
- Regulatory Limitations are dependent to a large extent on location and the aggressiveness of state and local government. Where government regulation has been increasing, any evaluation should assume that it will continue and likely become more assertive with resulting economic impacts on the cost of operation and required investment for the property.
- Capital Investment that is necessary to maintain and modify production, provide opportunities for an enhanced income stream, and assure efficient property management must be included in the evaluation. The latter includes provision for the orderly disposition of the property (AR&R costs).

# Economic Limit

At some point oil and/or gas production from a property does not generate enough revenue to cover all the costs associated with that production – the property ceases to make a profit and has reached the limit of production under the economic conditions imposed by the terms of the evaluation. Production may continue for a variety of management reasons, but the property has no economic value.

# Present Value

Petroleum property evaluations often extend over 10, 20, or 30 years, if not longer. For several reasons, the stream of income anticipated in these evaluations must be reduced to the equivalent value as of the effective date of the evaluation. A dollar to be received five years from today has less value than the receipt of that dollar today. There are many reasons for this; inflation, etc., are factors, but most reasons can be consolidated into “opportunity cost.” Even under the best of economic circumstances, the 5-year delay in receipt of income deprives the eventual recipient of the opportunity to otherwise invest that income and possibly receive further benefit. The present value concept leads to the issue of the discount rate to be used for oil property evaluation. While each company may have a methodology for derivation/selection of a discount rate, data from industry suggests that the financial cost-of-capital, in one variation or another, is the most common approach.

# 20

# LASTLY, A WORD ON "RESERVES"

Throughout the foregoing discussion of economic evaluation, reference is made, where necessary, to potential or ultimate recovery of crude oil and natural gas as a consequence of the evaluation process, not as an objective of the evaluation. There are several reasons for this deliberate usage of terminology. In most, if not all, petroleum evaluation texts the subject of "reserves" of oil or gas or other product tends to dominate the early discussion and imply that the objective of the process is to estimate reserves.

The premise of this text is that the primary objective of property evaluation is the estimation of the economic benefit to be derived from the property. To that end, the Jim Smith Rule is updated or augmented as:

*"We do not value reserves, we value income streams."*

However, in the interest of completeness, it may be useful to consider the concept of reserves and to examine the apparent necessity for definition, or possibly more nearly correct, description of the term. The American Petroleum Institute (API) is usually credited with publishing the first definition of (Proved) reserves in the 1950s. However, the Society of Petroleum Engineers (SPE) has taken the lead in attempting to establish the terms of reference for reserves, starting in 1964-65. The 1964 SPE definition was simple and concise, it actually fit on a postcard, and it defined Proved reserves as:

> *"The quantities of crude oil, natural gas and natural gas liquids which geological and engineering data demonstrate with reasonable certainty to be recoverable in the future from known oil and gas reservoirs under existing economic and operating conditions. They represent strictly technical judgments, and are not knowingly influenced by attitudes of conservatism or optimism."*
>
> *"When evaluating an individual property in an existing oil or gas field, the proved reserves within the framework of the above definition are those quantities indicated to be recoverable commercially from the subject property at current prices and costs, under existing regulatory practices, and with conventional methods and equipment."*

This is straightforward and served the purpose quite well. The definition did not get overly elaborate by trying to discuss every possible condition that might impact the calculation. It is sufficiently clear that the volume of reserves is a function of an array of physical and economic factors. Reserves are commonly expressed as a volume of oil or gas that is a function of certain existing and predicted economic conditions.

Further, reserves is an abstract concept, not unlike fair market value, that exists as a definition and depends upon the expectation of future production and sales of oil. Since different persons can have differing expectations for production and sales from a property, the reserves of that property can never be a fixed volume, but can and do change from observer to observer and from time to time, or both.

Given the clarity and easily understood nature of the SPE definition, it may have been a good idea to leave well enough alone. However, as noted by the Society of Petroleum Evaluation Engineers (SPEE):

> *"The continuing process of updating and revising reserve definitions is a reflection of both the changing nature of the petroleum industry with time and the difficulty of writing a comprehensive definition of reserves and guidelines that will serve the broad needs of industry without excess complexity."*

# In Summary

The SPE definition of 1964 is concise and direct while providing a perfectly functional description of reserves. The utility of the 1964 definition for property evaluation in 2023 (and beyond) is that the elaborate system of classes or categories of reserves is unnecessary. Classifications such as proved or probable are a function of likelihood of recovery of a volume of oil and/or gas production over a certain period of time. If the evaluation takes account of the of production by appropriate adjustment of the production projection and includes appropriate cost and investment allocations, there is no need for classification as proved or probable.

Reserves, as defined, are solely a function of economics. Oil & gas production from a property can occur, but unless that production is economic, it is not a reserve. Regardless of the size, shape, and duration of the production estimate, if there is no market for the production or if the revenue from the sale of the production is not sufficient to pay all operating costs, royalties, taxes, and regulatory imposts – and to repay investment – the income stream, and therefore the property, has no value.

Regardless of the simplicity or complexity of the economic evaluation(s) that may be conducted for a property, the value of the property rests solely in the present value of the income stream. In a spreadsheet calculation this value is usually in the bottom right-hand box. The ultimate recovery or reserves for that particular evaluation is found in the bottom left-hand box on a spreadsheet.

# APPENDIX A
# COMPARABLE SALES METHOD OF APPRAISAL AND PETROLEUM PROPERTIES

**Note:** Real Estate uses the term "*appraisal*" for the process of obtaining a value for a property rather than "*evaluation.*" The distinction is subtle but more than semantic.

Oil & gas evaluation borrows many concepts and procedures from real estate appraisal which is where most techniques were developed. The reference texts and journals are about real estate – land, homes, buildings. Oil & gas evaluation developed separately – from the ground up, so to speak – and can stand alone without reference to real estate practice. It is only when we move from evaluation to formal appraisal that we have to consult the real estate world. From the perspective of oil property appraisal, the concepts common to real estate are an exotic terrain stuck on the continental plate of oil & gas evaluation. It may occasionally be useful to refer to real estate methods or at least understand how they might relate to oil & gas.

According to the 11th Edition of the *Appraisal of Real Estate* (ARE):

*"Appraisers estimate property value with specific appraisal procedures which reflect three distinct methods of data analysis – cost, sales comparison, and income capitalization. One or more of these approaches are used in all estimations of value; the approaches employed depend on the type of property, the use of the appraisal, and the quality and quantity of the data available for analysis."*

Substitute "evaluator" for "appraiser" and the above applies to all who try to estimate the value of oil & gas properties. Many appraisers and most people working in the petroleum industry use the income approach virtually to the exclusion of the other two methods. There are good reasons for that situation and I do not anticipate anything said herein will change that. However, as professional appraisers, we should at least consider all three methods. For appraisers certified by peer groups such as the American Society of Appraisers (ASA) and the Appraisal Institute (AI), it is required. It is also required by the Uniform Standards of Professional Appraisal Practice (USPAP). This outline will examine the Comparative Sales Approach.

Real estate appraisers consider all three methods for three reasons:

1. The type of property and data available may not be amenable to evaluation using just one method.

As an example, consider a 10-year-old apartment building:

- An estimate of value could be based on the cost of reproducing or replacing the structure.
- An estimate could be based on obtaining data from recent sales of similar apartment buildings, making adjustments, and applying that value to the subject property.
- An estimate of value could be based on the expected income to be earned from continued operation of the property.

2. Using more than one method may account for characteristics and value considerations which might be missed using only one method. Assuming that values have been determined by each of the three methods, the evaluator must then reconcile the values to obtain *"a single value indication or a range of most probable values."*
3. The relative dependability and applicability of each approach are considered in reconciling the value indications into a final[134] estimate of defined value.

The Sales comparison approach is the process in which a value estimate is derived by analyzing the market for similar properties that have recently sold and comparing these properties to the subject property. This approach is commonly used in real estate appraisal for homes, land, farms, and commercial/industrial buildings. It is simple in concept and is usually thought to give a good estimate of value for the types of property that are appropriate to value by this method. The process has several steps. The appraiser researches the market for sales of properties with characteristics similar to the property being appraised, usually referenced as the **subject property.** After verifying the data regarding both the subject property and the property that was sold, the appraiser defines the similarities and differences between the comparable and subject property, and then makes adjustments to the value of the comparables to match the attributes of the subject property.

For the appraiser, most of the work in this method comes in the adjustments that have to be made to each of the comparables. These are termed **Elements of Comparison**. ARE lists ten such elements. These are:

- Real property rights conveyed,
- Financing terms,
- Conditions of sale,
- Market conditions,
- Expenditures prior to sale,
- Location,
- Physical characteristics,
- Economic characteristics,
- Use,

- Non-realty components of value.

Each of these elements may be broken down into sub-parts. **Physical characteristics** might include building size, lot size, quality of construction, architectural style, building materials, age, condition, functional ability, attractiveness, amenities, and on-site environmental conditions. Location might range from part of town or school district to whether the house is on a corner or in a cul-de-sac.

Assuming the appraiser has enough data, the adjustment process can start. This is essentially a process of determining which of the various characteristics of the comparable properties influence the value of each of those properties and by how much. It is hard work. All other things being equal, if one house has a pool and one does not, how much does the pool add to or detract from value? Since all other things are not equal, how much of the difference in value is due to the other elements of comparison? If there are only a few comparables, it may be necessary to use **paired-sales analysis** where one comparable to compared to another in an attempt to sort out the differences. If there is a sufficiently large and reliable database, statistical analysis, such as single or multiple regression, can be useful in defining differences. The relative reliability of the data and of the adjustments must be determined.

After sorting out the differences in value contributed by each of the property characteristics, the appraiser adjusts the value of each of the comparables using those elements that bring the comparable closest to the subject property. A common Unit of Comparison, such as $/sq. ft. or $/acre, is than applied to the subject property to obtain a value estimate.

This method apparently works well for the types of properties to which it is normally applied. In residential real estate all the necessary data is relatively easy to obtain from public records and from realtors' listings and databases. Banks and other sources provide financing and market information. If necessary, the appraiser can go to each property and accumulate data directly – measure the house, the rooms, check the heating, the roof, the plumbing, etc. He can talk to the buyer, the seller, real estate agents, the finance company, and escrow agents. This could be done in our condo development just by calling the realtor who has sold most of the units over the last 15 years.

# Comparable Sales for Oil Properties

Why is comparable sales not used to value oil properties? There has been only one serious effort to apply sales comparison to oil properties, but it never really caught on.

It should be simple to cite several reasons why the comparable sales method is not used to value oil properties. They are definitive and would not be a surprise.

But let us be positive and examine the process of valuing oil properties using a correct sales comparison approach.

Assume that the subject property is a single oil-producing property with all the necessary components of production. It is then necessary to locate as many recent sales of properties with characteristics similar to the subject as feasible. Recent sales are important; otherwise you have to make more complex adjustments for market and financial conditions. The more similar the properties, the fewer and smaller the adjustments. Already there is a problem because oil properties do not sell very often – certainly, not in the numbers and with the frequency of houses, condos, or land. The number of comparables is limited, particularly when considered in a recent time frame.

Assuming that a few sales are found that are timely – what is known about the properties? How can the sales price be determined? There are no realtors, no multiple listings, no public reporting. Even public companies only have to report transactions that are material. Property tax records are confidential. Working backwards from deed stamps is approximate at best. Some data regarding significant transactions may be published in the newspaper, the *Oil & Gas Journal* (OGJ), or other publication. However, published prices for major transactions are rarely sufficiently detailed to know what was really paid, what other non-cash consideration was involved, or how the transaction was financed.

For the sake of argument, assume that reliable purchase prices can be obtained for several sales and that some information about the properties is known. It is now necessary to adjust the comparable properties to the subject property.

1. What are the similarities and differences? The list here could be endless. Think of all the physical differences that could occur between two oil properties.
2. Think of all the operational differences that could occur. Third, list all the economic differences.
3. List all the circumstances of the sale that could influence the value.

I came up with a couple of dozen physical differences starting with location, field, reservoir, oil gravity, depth, number of wells, etc. If the four lists are laid out side by side, many of the items are found to be related and some are duplicates – oil gravity influences price, etc. Fine tuning the lists provides a matrix of physical and other factors that could influence the value of a property.

The next step is to determine the degree of confidence that can be assigned to the data. Where did the data come from or where did the data source get it? Now determine how much influence each factor in the matrix has on the sales price of the comparable property. How much of the sales price is due to Comparable #1 having a sandstone reservoir while Comparable #2 produces from limestone? How much difference is due to Comparable #3 having a bottom-water drive while Comparable # 1 is a steamflood? How much difference is due to Comparable #1 having 13° API oil while Comparable #3 is 21° API?

Comparable sales analysis requires specific adjustments for real differences. Waving your hands and saying #2 is better than #3 by some imaginative criteria is not enough. This is where that regression analysis comes into play. There must be definable and supportable differences in value which can be related to recognizable characteristics of the properties.

For the sake of discussion, assume that a rigorous comparable sales analysis has been completed and you have adjusted your comp to the subject property. Now it is necessary to define a Unit of Comparison. The old \$/Bbl or \$/BOE is often mentioned here, but a little thought would reveal a few problems. The \$/ Bbl unit worked well enough 30-40 years ago when oil prices were relatively stable; most future production was classed as proved and predictable, and income valuations were difficult. Using \$/Bbl would require a definition of how many dollars and what kind of barrels. Just Bbl of oil or a combination of crude oil and natural gas defined as barrels of oil equivalent (BOE). Is there enough information about the potential recoverable oil of each comparable? So, would not a unit of comparison related to future income be more useful? But then you would have to know the anticipated future income, and here again the required data is missing.

Assuming that a useful unit of comparison could be derived, the subject property could be valued by applying the appropriate unit. If the unit is \$/acre and the subject has 640 acres, then the value is 640 x \$/acre.

It should be apparent that comparable sales cannot be applied to oil properties, because:

- there are not enough sales.
- there is not enough available data.
- there is no accepted unit of comparison.
- oil properties are acquired for the income they produce, not for the amenities of the property.

# Serious Attempts

As noted above, there have been attempts to adapt Comparable Sales to oil properties. The best of these was by Bill Strevig in the late 1980s (SPE#18907). This was a form of comparable sales method which started from a \$/BOE for each comp and/or for a group of sales and made adjustments to the \$/BOE before applying it to the subject property. There were at least 19 specific criteria for which adjustments should be made and a direction to make as many others as necessary. This approach never really caught on because, aside from investment bankers and a few consultants, very few people had the data resources to make it work and those resources were almost entirely income approach evaluations. Without the knowledge provided by the income evaluations, there is no starting point and no source for making adjustments. For all its difficulty, the Strevig method was at least a reasonable representation of a market sales approach. It could be done, but it had to be right.

## Not-So-Serious Attempts

The same cannot be said of other attempts to invoke the comparable sales method. For the most part, these mutations start from an income approach evaluation, derive a convenient ratio (usually $/Bbl or $/BOE), fail to identify physical, operational, or economic differences between the comparable and subject, and ignore the necessity of making specific adjustments between each comparable for identified differences. These methods rely instead on overall adjustments based on differences between the income projections for each of the properties.

# APPENDIX B PRIMARY PRODUCTION IN PETROLEUM RESERVOIRS

At the risk of repetition, right about here in most evaluation texts, the narrative transitions to a lengthy discussion of reservoir engineering and the various methods of estimating future recovery based on the geology of the reservoir, pressure and fluid flow characteristics, well design, and production development. There are numerous texts that cover those subjects in great detail. In addition to petroleum engineering and geology books and papers, many companies have developed computer-based systems to evaluate oil & gas reservoirs including the estimation of future production; some of these are commercially available to evaluators. Given that those sources exist and can be readily accessed, this discussion will not repeat the information in any depth but, instead, borrows extensively from one or more of those texts. The reason is rather simple: these were the texts from which most petroleum engineers learned the fundamentals of their profession and, judging by the content of more recent texts, in some cases, the subject has not changed much over the decades. Where appropriate, attribution and references to the sources will be provided, but the following sources are the most relevant.

1. *Petroleum Production Handbook*, vol. II, Reservoir Engineering, edited by Thomas C. Frick, McGraw-Hill, 1962; hereinafter referenced as **PPH** with Chapter and page noted as 37-10, respectively.
2. *Petroleum Engineering Handbook*, edited by Howard B. Bradley, Society of Petroleum Engineers, Third Printing, 1987; hereinafter referenced as **PEH** with Chapter and page noted as 37-10, respectively. PEH is largely a reprint of PPH with some new material.
3. *Applied Petroleum Reservoir Engineering*, Craft, B.C. and Hawkins, M.F., 1959 (there are later editions), Prentice-Hall; hereinafter referenced as **C&H** with standard page numbering.
4. *Petroleum Reservoir Engineering*, Amyx, James W., Bass, Daniel M., and Whiting, Robert L., 1960, McGraw-Hill; hereinafter referenced as **AB&W** with standard page numbering.

Each of the above sources exists in later editions where, for the most part, the only significant changes are the addition of color and the price of the book.

# Primary production

Primary production uses the natural reservoir energy of free and/or dissolved gas, encroaching water, or gravity, separately or collectively, as the recovery mechanism. The following is a summary of some approaches to reservoir analysis as a means of deriving a projection of future performance. There are several methods available for estimation of potential recovery which can be used alone or in concert. The use of the methods depends, to a degree, on the stage of life of production from the property and the amount and quality of data that is available. These methods can be generally grouped as: Volumetric, Material Balance, and Production Performance. Each method has its attributes and drawbacks.

# Volumetric Methods

Volumetric Methods are most commonly used early in the life of a property, usually before significant production has occurred. This is a subjective criterion; volumetric methods can also be used long after a property has reached maturity, as production and reservoir data has accumulated that might modify previously estimated volumes, or to study particular segments of the reservoir for further development. Volumetric methods attempt to determine the amount of oil-and/or gas-in-place, and the potential recoverable volume of hydrocarbons based on certain reservoir characteristics. The method requires a knowledge of both the possible size of the reservoir in terms of rock volume (measured in acre-ft), and the physical properties of the reservoir rock(s) and fluid(s). The method assumes that the reservoir rock has a certain volume of void space, measured as *porosity*, that the pore space is interconnected (*permeability*), and that the void space contains fluids (oil, water, gas). The volume of original oil-in-place (OOIP), in a given segment of the reservoir, is approximately:

$$OOIP = V_{o=}\left(V\varnothing\right)S_{h}, \text{ Bbl STO}$$

where: V = Estimated volume of reservoir measured in acre-feet

$\phi$ = Porosity – % of void space in V

$S_h$ = Hydrocarbon saturation as a % of fluid content

= 1 – $S_w$, where $S_w$ is the water saturation as a % of fluid content, or

= $S_o + S_g$ if there is a free gas cap

For practical use, the equation is:

$$OOIP = V_o = \frac{7758V\left(\varnothing\right)\left(1-S_w\right)}{B_o}, \text{ Bbl STO} \qquad \text{[PPH 37-10(7)]}$$

where (7758) is the conversion from acre-feet to barrels and ($B_o$) is the oil formation volume factor used to convert the fluid volume at reservoir pressure and temperature to surface stock tank barrels.

In free gas reservoirs or in the gas cap of oil reservoirs, the volume of gas-in-place is estimated by:

$$OGIP = V_g = \frac{43560V_g(\varnothing)(1-S_w)}{B_{gi}}, \text{ SCF free gas} \qquad \text{[PPH 37-10(6)\}}$$

where (43560) is the conversion from acre-feet to cubic feet of reservoir volume and ($B_{gi}$) is the gas formation volume factor used to convert the gas volume at reservoir temperature and pressure to surface conditions. The calculation is of Standard Cubic Feet (SCF) of gas at initial ($B_{gi}$) reservoir conditions. The gas formation volume factor ($B_g$) may be estimated for various combinations of pressure, temperature, and gas gravity from published tables.

The volume of dissolved gas in an oil reservoir can be estimated by:

$$OGIP = V_g = \frac{775V(\varnothing)(1-S_w)R_s}{Bo}, \text{ SCF free gas} \qquad \text{[PPH 37-10(8)]}$$

where ($R_s$) is the estimated volume of dissolved gas (SCF) per barrel of oil at reservoir conditions.

As is apparent from the equations, the volumetric method has certain data requirements:

- Reservoir Volume
    - Reasonably accurate mapping of gross and net sand (the latter should be smaller than the former)
    - Determination of oil-water and gas-oil contacts
    - Calculation of approximate reservoir volume (acre-ft)
- Reservoir Properties
    - Estimate porosity – from logs, cores, or both
    - Estimate gas and/or water saturation – from logs, cores, or both
    - Estimate residual ($S_o$) and ($S_w$) – from core tests
- Fluid Properties
    - Determine ($B_o$) at initial and abandonment conditions from PVT analysis
    - Determine ($B_g$) at initial and abandonment conditions from PVT analysis

The volumetric method can be subject to considerable error because it is commonly used to evaluate a property when data may be limited and it requires the estimation of reservoir rock and fluid properties and reservoir volumes from spot (borehole) measurements of those properties (which may be altered by the drilling and sampling process), which are then extrapolated to the entire reservoir. Reservoir volume and composition is estimated using a *zone thickness* measured at one or more points in the reservoir. While the advent of increasingly sophisticated 3-D seismic and other geophysical methods has made possible improved delineation of potential reservoir structures, the areal extent and thickness of the reservoir is rarely known until a large number of wells are drilled. Porosity and water saturation are obtained from well logs and/or core samples which are extracted from a small volume of the reservoir and which, under the best of circumstances, only approximate the conditions in the reservoir. Gas saturations ($R_s$) are derived from laboratory analysis of representative samples. Even though techniques of core analysis and well log measurement and analysis have advanced substantially, the volumetric method, particularly in the early life of the reservoir, remains only a gross estimate of hydrocarbons-in-place.

## Recovery Factor

The Volumetric Method estimates the volume of oil (OOIP) and/or gas (OGIP) available in the reservoir using the best knowledge and under the circumstances at the time of the estimate. While many old oil fields have recovered a substantial portion of the original oil-in-place, often over decades and, more times than not, employing secondary and tertiary recovery methods, all the hydrocarbons in the reservoir can never be produced. Conversion of a volumetric oil-in-place to recoverable volume requires the use of a **recovery factor** which is an estimate of the portion of the original volume that can be expected to be produced under given reservoir conditions. The recovery factor can be a unit recovery (Bbl or MCF per acre/ft), a percentage of OOIP, or both. It can be measured at reservoir conditions but is more commonly (and usefully) derived at surface conditions.

Recovery factors can be derived from the performance of similar reservoirs, from laboratory analysis of cores, or from computer simulation of the anticipated reservoir performance. If data is available, comparison to similar reservoirs which have been sufficiently developed to allow an estimate of future primary recovery is likely the best source of information. As noted earlier, there are a number of caveats to this approach including, but not limited to, the need to assume that the property being evaluated would be developed and operated in the same manner as the property used as a comparison. Many older textbooks and engineering/geologic studies discuss recovery factors for various fields, operating conditions, and location, often including extensive data collections and providing tables of observed and calculated recovery factors from certain fields and reservoirs.

# Laboratory Core and Fluid Analysis

Laboratory core and fluid analysis is commonly coupled with computer simulation to model development and production from the reservoir. Core and fluids analysis provides the data for the computer model. This combined model approach is time-consuming (and could be expensive), but it has at least two advantages.

First, the models can calculate potential recoveries under varying reservoir rock and fluid conditions, assumed production schedule and methods, and operational regimes. As with all models, the more reliable the data, the better. Second, there is no need to evaluate similar reservoirs to determine if there is sufficient commonality to allow comparison. However, reference to data from similar reservoirs could provide valuable information for comparison with the model results.

Absent a computer model, a recovery factor can be calculated assuming information is available from reservoir fluids and core analysis.

Producible Volume = N = OOIP x Recovery factor, or

Rf = Recovery Factor = N/OOIP

For solution gas drive or depletion drive reservoirs:

$$N = 7758(\varnothing)\left[\frac{(1-S_w)}{B_{oi}} - \frac{(1-S_w-S_{gr})}{B_{oa}}\right], \text{ SBT/Acre-ft} \qquad \text{[PPH 37-13(19), PEH 40-8(14)]}$$

where potential recovery is the difference between oil-in-place (OOIP) at initial reservoir conditions and remaining OIP at assumed abandonment conditions.

In a water drive reservoir:

$$N = 7758(\varnothing)\left[\frac{1-S_w}{B_{oi}} - \frac{S_{or}}{B_{oa}}\right], \text{ STB/Acre-ft} \qquad \text{[PPH 37-25(35), PEH 40-16(31)]}$$

Note that these equations require reservoir fluids information ($S_w$, $S_{oi}$, $B_{oi}$, $B_{oa}$, $S_{or}$) that usually only comes from laboratory core and fluids analysis. Estimation of potential recoverable volumes of oil and/or gas is an important step, but the valuation of the property requires a reliable estimate of the future production performance. It is still necessary to determine the production rates and the duration of production that will obtain the estimated volume.

The volumetric methodology, even when used in concert with a range of potential recovery factors, produces only a volume of potential recovery but provides no information as to the schedule of recovery.

# Projections from Volumetrics

Developing a production projection based solely on an estimate of oil-in-place, presumably subject to a recovery factor, can be difficult unless supported by a production rate model study or data from similar properties. The observed tendency is to overestimate the production capacity, well drilling rates, and surface facility efficiency of the property. It is rare but cases have occurred when the evaluator built a production schedule designed to recover the estimated recoverable oil (or gas) without reference to likely performance derived from reservoir analysis or models from other sources.

In this instance it is essential that some effort be made to derive production/ development models from similar properties and operations. The failure to make use of practical models can be illustrated by an example. In 2000, an evaluation of a proposed new field development offshore California included volumetric estimates of OOIP and recoverable oil for a property where there was limited drilling (including two dry holes) and testing and no measurable production. The property was located between two fully developed fields with established production histories. Data for those fields was available from public and private sources. However, the evaluation chose to create a production projection that ignored the offset fields performance and instead built an artificial projection designed to recover the calculated volumetric amount of oil & gas. Comparison of the estimated production for the new property to the performance of the offsetting properties indicated that the artificial projection was significantly optimistic. The property was never developed.

# Material Balance

The material balance is a more complex method of estimating original oil-in-place and potential recovery, but it has the advantage of providing an estimate of production over time under a given set of conditions. The method has several forms (as developed by researchers over time) and requires both an extensive Pressure-Volume-Temperature (PVT) analysis of reservoir fluids and a reasonably accurate model of the production mechanism for the reservoir. The latter requires that some production (5-10% of ultimate recovery) occur before the method can be applied. The material balance is not a substitute for the volumetric method but can be used along with the volumetric method later in the life of the property and is often used to obtain and/or refine a recovery factor for volumetric calculations. The material balance could also be used if reliable pressure performance data can be obtained from a reservoir with similar rock and fluid properties.

Material balance is a practical application of the principles of Conservation of Mass and Energy and is designed to simulate the balance of the withdrawals from a reservoir with resulting changes in volume of the original reservoir fluids and the possible influx of additional fluids. The several forms in common use attempt to model reservoir performance over time by equating the expansion of reservoir fluids (gas and/or water), as pressure decline occurs during production, to the change in void-space of the reservoir caused by withdrawal of oil, gas, and water (less any water influx). The method has been modified, adapted, and, in a few cases, simplified by many authors. The method proposes an iterative solution of sequential pressure reductions caused by reservoir fluids production where values for each successive step are taken from PVT analysis of the reservoir fluids. [PPH 37-10, PEH 40-6]

The Material balance method requires substantially more data and calculation than the volumetric method; however, where it can be applied, it usually provides an incrementally more reliable result since it incorporates the dynamic changes in reservoir fluids rather than being a static calculation. Given the complexity of the calculations, the adaptation of the method to computer analysis seems obvious. The result, however, is still only an estimate of total recoverable volume. As noted above for the volumetric method, the valuation of the property requires a reliable estimate of the future performance that relates production to time.

The comprehensive material balance general equation is:

$$N = \frac{N_p\left[B_t + 0.1781 B_g\left(R_p - R_{si}\right)\right] - \left(W_e - W_p\right)}{B_{oi}\left[m\left(\frac{B_g}{B_{gi}}\right) + \frac{B_t}{B_{oi}} - \left(m+1\right)\left(1 - \frac{\Delta p\left(Cf + S_w C_w\right)}{1 - S_w}\right)\right]}$$ [PPH 37-12(10), PEH 40-6(5)]

where N is Original Oil-in-Place (OOIP) which, by definition, must remain the same through each step of pressure decline. A casual review of the general equation suggests a couple of important considerations before leaping into a material balance analysis.

1. The general equation assumes that there is a free gas cap ($R_p$) and an active water drive ($W_e$).
2. The general equation assumes that all fluid and rock components of the reservoir react to production. This is rational and should be expected if conservation of mass and energy are to be attained. However, reservoir behavior may suggest that one or more components of the equation are either not significant or not measurable, such as water influx, or the reservoir may be above the bubble point. In those circumstances, certain components may be deleted from the equation which produces modifications of the calculation.

As an example, if reservoir rock compressibility ($C_f$) and connate water compressibility ($C_w$) can be neglected, then the equation can be reduced to:

$$N = \frac{N_p\left[B_t + 0.1781 B_g\left(R_p - R_{si}\right)\right] - \left(W_e - W_p\right)}{mB_{oi}\left(\frac{B_g}{B_{gi}}\right) + \left(B_t - B_{oi}\right)}$$ [PPH 37-11(11), PEH 40-6(6)]

Water influx ($W_e$) is rarely known or measurable; however, where water influx occurs, calculated values of (N) represented by a new value, ($N_i$), over time would increase. (N) could be estimated by plotting ($N_i$) vs time and extrapolating back to t = 0 where $N_t$ (t = 0) = N. This approach includes a significant potential for error if the plotted points are not a straight line

If the reservoir has no discernable water drive, then water influx ($W_e$) can be deleted, resulting in:

$$N = \frac{N_p\left[B_t + 0.1781B_g\left(R_p - R_{si}\right)\right] + W_p}{mB_{oi}\left(\frac{B_g}{B_{gi}} - 1\right) + \left(B_t - B_{oi}\right))}$$ [PPH 37-11(12), PEH 40-7(7)]

Or, if the reservoir is initially under-saturated, with no free gas, below the bubble point, and no water influx, the equation can be further simplified to:

$$N = \frac{N_p\left[B_t + 0.1781B_g\left(R_p - R_{si}\right)\right] + W_p}{\left(B_t - B_{oi}\right))}$$ [PPH 37-11(16), PEH 40-7(9)]

3. There are no pressure terms included in the equations. Use of the material balance requires that a series of pressure changes and estimated fluid production volumes be scheduled, which are then related to an estimate of changes in fluid properties brought about by the pressure-volume reductions. Variations in the material balance equations can be constructed depending on the particular circumstances of the reservoir.
4. The material balance is a complex method that relies on access to and analysis of the physical properties of reservoir fluids (oil, water, and gas) that may or may not be available. The result can be compromised by incomplete knowledge of the reservoir, constituent fluids, etc. Laboratory analysis of oil, water, and gas properties as a function of changes in pressure must be done prior to using the material balance method. Fortunately, some of this work has been done for many reservoirs and fluid samples and has been published as data tables or as graphs which allow access to fluid properties and in some cases provide estimates of potential recovery without having to go through the material balance calculation.
5. The material balance equations and the methodology for obtaining a satisfactory result can be time-consuming and tedious, particularly if the intent is to define potential production for a comparatively small property. The adaption of the method(s) to computer programming and application on even a small computational system can alleviate a considerable amount of both tedium and cost, which can then be applied to enhancing the quantity and quality of fluid properties response.

By using the pressure-production history of the reservoir to obtain a relation of Np (production) to Δp (change in reservoir pressure), a projection of Np/Δp can be used as the basis for an estimate of future production to the point of pressure depletion. At that point, ultimate primary recovery is obtained and:

$$\text{Recovery Factor} = R_f = Np/N$$

# Unit Recovery

*"The unit-recovery factor is a theoretically possible ultimate recovery in stock-tank barrels from a homogeneous unit volume of 1 acre-ft of pay produced by a given mechanism under ideal conditions. The unit-recovery for a saturated depletion-type reservoir is equal to the stock-tank oil initially in place in barrels per acre-ft at pressure $P_i$ minus the residual stock-tank oil under abandonment pressure. By difference the unit recovery by depletion or solution-gas drive is:*

$$7758\varphi = \left( \frac{1 - S_w}{B_{oi}} - \frac{1 - S_w - S_{gr}}{B_{oa}} \right) \qquad \text{[PPH 37-13(19), PEH 40-8(14)]}$$

*The key to computation of unit recovery by means of this formula is an estimate of the residual gas saturation $S_{gr}$ at abandonment time."*

*"The unit recovery for a water-drive reservoir is equal to the stock-tank oil originally in place in barrels per acre-foot minus the residual stock-tank oil at abandonment time. By difference the unit recovery by water drive is therefore:*

$$7758\varphi = \left( \frac{1 - S_w}{B_{oi}} - \frac{S_w}{B_{oa}} \right) \qquad \text{[PPH 37-25(35), PEH 40-16(31)]}$$

*The residual oil saturation $S_{or}$ at abandonment time may be found by actually submitting cores in the laboratory under simulated reservoir conditions to flooding by water."*

Several authors have presented variations in the calculation procedure to obtain material balance recovery factors. All have pursued the objective of solving the material balance calculations to obtain a usable recovery factor in a practical solution. While, as noted, the calculation procedure is eminently applicable to computer simulation, a knowledge of the underlying methodology can be useful.

## Muskat's Method

Muskat's method calculates the change in oil saturation with change in pressure as production occurs. The approach assumes uniform oil saturation (no gas segregation) and relatively low permeability, requires calculation using small pressure increments, and therefore, requires detailed analysis of reservoir rock and fluid properties as a function of reservoir pressure.

The basic equation for the Muskat method is described by Arps [PPH 37-14], *"If the actual relationships between pressure and oil-formation volume factor* $B_o$*, gas-formation volume factor* $B_g$*, gas-solubility factor* $R_s$*, oil viscosity* $\mu_o$*, and gas viscosity* $\mu_g$ *are available from a PVT analysis of the reservoir fluids, and if the relationship between relative permeability ratio* $K_{rg}/K_{ro}$ *and the total liquid saturation is known for the reservoir rock under consideration, then the unit recovery by depletion can be arrived at by a stepwise computation of the desaturation history directly from following depletion equation in differential form:"* [PPH 37-14(20), PEH 40-9(15)]

$$\frac{\Delta S_o}{\Delta p} = \frac{S_o\left[\frac{B_g dR_s}{B_o dp}\right] + \left(1 - S_o - S_w\right)\left[B_g \frac{d\left(1/B_g\right)}{dp}\right] + S_o\left[\frac{\mu k_{rg} dB_o}{\mu k_{ro} B_o dp}\right]}{1 + \frac{\mu k_{rq}}{\mu k_{ro}}}$$

The creation of tables and graphs of some of the equation terms as functions of pressure changes can facilitate the procedure. Arps and others suggest streamlining the process by breaking down the broad equation into discrete components and reducing them to a graphical model.

$$\left(\frac{B_g}{B_o}\frac{dR_s}{dp}\right), \left[B_g\left(\frac{d\left(1/B_g\right)}{dp}\right)\right], \frac{\mu_o}{\mu_g}, \left(\frac{1}{B_o}\frac{dB_o}{dp}\right) \qquad \text{[PPH 37-14]}$$

*"and the relative permeability ratio* $k_{rg}/k_{ro}$*, which is a function of total liquid saturation,* $S_t$*, only."*

Where, at depletion, $S_t = S_{oi} - \Delta S_o$, and

$$R_f = \frac{S_{oi} - S_{oa}}{S_{oi}}, \%$$

The results are converted into cumulative recovery per acre-foot by:

$$N_p = 7758(\phi) = \left(\frac{1 - S_w}{B_{oi}} - \frac{S_o}{B_o}\right), \text{STB/acre-ft} \qquad \text{[PPH 37-14(21), PEH 40-9(16)]}$$

Cumulative recovery as a percentage of OOIP can be determined by:

$$\frac{N_p}{N} = 1 - \left(\frac{S_o}{1 - S_w}\right)\left(\frac{B_{oi}}{B_o}\right), \% \qquad \text{[PPH 37-14(22), PEH 40-9(17)]}$$

and gas-oil ratio performance by:

$$R = R_s + 5.615\left[\frac{B_0}{B_g}\frac{\mu_0}{\mu_g}\frac{k_{rg}}{k_{ro}}\right], \text{SCF/STB} \qquad \text{[PPH 37-14(23), PEH 40-9(18)]}$$

and a relative production rate model:

$$q_0 = \frac{k_0}{k_{oi}}\frac{\mu_{oi}}{\mu_0}\frac{p}{p_i}q_{oi}, \text{Bbl/Day} \qquad \text{[PPH 37-14(24), PEH 40-9(19)]}$$

*"It should be stressed that this method* [Muskat] *is based on the assumption of uniform oil saturation in the whole reservoir and that the solution will therefore break down when there is appreciable gas segregation in the formation. It is therefore applicable only when permeabilities are relatively low."* Arps [PPH 37-14]

## Tarner's Method

Tarner's method, for use in solution gas reservoirs below the bubble point (BP), requires a simultaneous solution of the material balance equation and the instantaneous gas-oil ratio equation. The procedure is to calculate the cumulative oil (Np) and gas (Gp) for a pressure drop ($p_1 - p_2$) as follows:

1. $N_p = 0$ at bubble point

2. $$\left((G_p)_2 = (N_p)_2 (R_p)_2 = N\left[(R_{si} - R_s) - 5.615\left[\frac{B_{oi} - B_0}{B_g}\right]\right] - (N_p)_2\left[5.615\frac{B_o}{B_g} - R_s\right]\right)$$

   [PPH 37-15(25), PEH 40-10(20)]

3. $$S_{t2} = S_w + (1 - S_w)\ \frac{B_0}{B_{oi}}\left[1 - \frac{N_{p2}}{N}\right] \qquad \text{[PPH 37-15(26), PEH 40-10(21)]}$$

4. Determine $k_{ro}/k_{ro}$ at $(S_t)_2$ :

$$R_2 = R_s + 5.615\left(\frac{B_0}{B_g}\frac{\mu_0}{\mu_g}\frac{k_{rg}}{k_{r0}}\right) \text{ at } p_2 \qquad \text{[PPH 37-15(27), PEH 40-10(22)]}$$

5. Compute the cumulative produced gas at pressure $p_2$:

$$(G_p)_2 = (G_p)_1 + \left(\frac{R_1 + R_2}{2}\right)\left[(N_p)_2 - (N_p)_1\right]$$

   at $p_2$ [PPH 37-15(28),PEH 40-10(23)]
   Where ($R_1$) represents the instantaneous gas-oil ratio computed previously at pressure ($P_1$).

6. Make three or more good estimates of $(N_p)_2$ and the corresponding $(G_p)_2$ from Steps 2 and 5. Plot $(N_p)_2$ vs. $(G_p)_2$ for Step 2 and for Step 5 and take the intersection of the curves as satisfying both equations.

Published tables of recovery factors, calculated using the Muskat and Tarner methods, can be used where no detailed data regarding reservoir fluids or rock properties is available. [PPH 37-16, Table 37-4 & PEH 40-10, Table 40.4]

These two methods and numerous other equally useful approaches (see references to Babson, and Tracy) to material balance calculations share a couple of things in common. First, they are relatively old (or vintage, if you prefer) methods of reservoir evaluation developed in the 1930s and 40s. Fortunately, as noted earlier, the physics and chemistry of reservoir fluid dynamics are well known and do not change; our understanding of and ability to model those dynamics may improve but the basics remain the same. Second, in the period before the IBM desktop computer and Sharp electronic calculator, the Muskat, Tarner, Babson, and other methods were designed to be done (and were done) by incremental calculations using slide rules and/or mechanical calculators with the results posted on 20-column accounting sheets. They were time-consuming but, depending on data quality and quantity, could provide reasonable results. It should be obvious, however, that repetitious iterative calculations are best done on modern desktop and laptop computers. All these methods, as well as many new adaptations, have been built to be used in Excel and other formats.

# Non-Associated Gas Reservoirs

The estimation of producible volumes for non-associated gas reservoirs is less complex than for oil reservoirs if only because the fluid dynamics are simpler and similar field performance is more directly applicable. The basic principles of volumetric gas-in-place and recovery factor are the same as for crude oil.

$$OGIP = G = \frac{43560 V_g (\phi)(1 - S_w)}{B_g}, \text{ SCF} \qquad \text{[PPH 37-32, PEH 40-23]}$$

*"The unit-recovery factor (the theoretically possible ultimate recovery from a homogeneous unit volume of pay under ideal conditions) for a dry-gas reservoir without water influx is equal to the gas initially in place at pressure pi, minus the gas remaining under abandonment pressure pa, both expressed in standard cubic feet per acre-foot of sand."* Jan Arps, PPH 37-37

At initial conditions: $$G_i = \frac{43560 \omega (1 - S_w)}{B_{gi}}, \text{ SCF/acre-ft} \qquad \text{[PPH 37-37]}$$

At abandonment condition: $$G_a = \frac{43560 \omega (1 - S_w)}{B_{ga}}, \text{ SCF/acre-ft} \qquad \text{[PPH 37-37]}$$

*"By difference the unit recovery for a dry-gas reservoir is....:"* [Arps 37-37]

$$\Delta G_p = 43560(\phi)(1 - S_w)\left[\frac{1}{B_{gi}} - \frac{1}{B_{ga}}\right], \text{ SCF/acre-ft}$$ [PPH 37-37(41), PEH 40-24(37)]

The gas formation volume factor ($B_g$) at abandonment ($B_{ga}$) is calculated at abandonment pressure ($P_a$), which can be based on pipeline pressure or the minimum pressure to which the reservoir can be reduced for the conditions of reservoir depth, tubing size, or other constraint(s).

*"The general material balance equation can be reduced to a form that will permit the calculation of the initial gas in place. Since there is no liquid production concerned in this evaluation, the oil in place N is zero, the solution-gas production is zero, and the gas-cap production $G_{pc}$ is the total gas production $G_p$. For a small pressure difference the change in gas volume is so large in comparison with the changes in rock and connate-water volumes that the change in water and rock volume can be treated as zero. For simplification, it is assumed that the water influx, water production, and water and gas injection are all zero."* (AB&W, p. 566)

Inserting the assumptions into the general material balance equation becomes:

$$G = 5.61(B_{gc} - B_{gi}) = (G_{pc}B_{gc} / 5.61)$$ (AB&W, p. 573)

After rearranging:

$$G = \frac{(G_{pc}B_{gc})}{(B_{gc} - B_{gi})} = \frac{G_{pc}}{(1 - (B_{gi} / B_{gc}))}$$ (AB&W 8-30)

The gas formation volume factor Bg is calculated where (Z) is the gas compressibility: (AB&W, pp. 236–237)

$$B_{gc} = \frac{0.00504TZ}{P}$$

Since reservoir temperature is considered to be constant, the equation then becomes:

$$G = G_{pc}\left[\frac{1}{1 - (P / Z)(Z_i / P_i)}\right]$$ (AB&W 8-31)

which can be solved with limited production and laboratory data. *"The equation can be rearranged to express the ratio of reservoir pressure and its compressibility factor as a linear function of the cumulative gas production."* (AB&W p. 566)

$$\frac{P}{Z} = \left(-\frac{P_i}{GZ_i}\right)G_{pc} + \frac{P_i}{Z_i}$$ (AB&W 8-32)

The equation "*indicates that the gas reserves can be estimated from a graphical plot of P/Z and the cumulative gas production Gpc. Extrapolation of the resulting curve to P=0 represents the initial gas in place, whereas extrapolation to the appropriate P/Z ratio corresponding to abandonment indicates the gas reserves.*" (AB&W, p. 566)

# Gravity Drainage

"*Gravity drainage is the self-propulsion of oil downward in the reservoir rock. Under favorable conditions it has been found to effect recoveries of 60 per cent of the oil in place, which is comparable with or exceeding the recoveries normally obtained by water drive. Gravity is an ever-present force in oil fields that will drain oil from reservoir rock from higher to lower levels wherever it is not overcome by encroaching edge water or expanding gas.... A thorough discussion of the many aspects of gravity drainage will be found in a classic paper by Lewis. Several authors have attempted to formulate gravity drainage analytically, but the relationships are quite complicated and not readily adaptable to practical field problems.*" Arps [PPH 37-23] quoting, in part, from the seminal paper by Lewis, *Gravity Drainage in Oil Fields,* Lewis, James O., AIME, October 1942.

"*Gravity has always been recognized as one of the three main recovery agents but generally it has been dismissed as relatively unimportant, consequently study of it has been neglected.... Field evidence shows that in some reservoirs gravity has been effective and has resulted in recoveries equal to or exceeding what could be expected from water displacement.*" (Lewis)

Gravity drainage really does not get much respect. All the attention is given to solution-gas and water-drive production and the vast array of studies, equations, and analyses related thereto. Arps offers two good examples of gravity drainage: Lakeview (which blew-out as a gusher) in Kern County, California, and the Oklahoma City Wilcox reservoir in Oklahoma. Later publications could probably add more examples.

It is likely that gravity drainage, because it is ubiquitous, is acting in many more reservoirs than are reported but the effect is masked by more active or observable production mechanisms. As noted by Arps, isolating the gravity effect is not easy but particularly where reservoir rock and fluid conditions cooperate, the additional energy added by gravity should be expected even if it cannot be directly quantified. The effect is most likely to occur in steeply dipping reservoirs. An example is the Monroe Swell field in Monterey County, California.

# APPENDIX C DERIVATION OF A BEFORE FEDERAL INCOME TAX COST-OF-CAPITAL DISCOUNT RATE

Derivation of the Weighted Average Cost-of-Capital (WACC) has been the subject of uncounted financial texts along with hundreds of published and unpublished papers in finance, appraisal, and business management, as well as at least one Nobel Prize for work on an issue in the cost-of-equity. Those collected works are summarized in the following paragraphs. The Before Federal Income Tax (BFIT) WACC is a special purpose variation that requires some additional discussion. Two significant publications are recommended as useful sources of BFIT WACC discussion and derivation of discount rates directly related to petroleum property evaluation.

The first is *"A Manual for Discounting Oil and Gas Income"* published by the Texas Comptroller of Public Accounts ("the Texas Manual").[135] The Texas Manual has been prepared since the mid-1980s by the Property Tax Assistance Division (PTAD) of the Comptroller's office in response to state law to provide information and direction to Texas property tax appraisal districts in valuing oil & gas properties using Discounted Cash Flow (DCF). The Manual (last updated in June, 2021) includes an extended discussion of the DCF methodology and the use of WACC as a source of discount rates. Further, it presents a detailed discussion of the calculation of a WACC-based discount rate including examples of the WACC calculation.

The second publication is Assessors Handbook 566, *Assessment of Petroleum Properties* published by the California State Board of Equalization (CSBE) to provide information and direction to County Assessors in the appraisal of oil & gas properties for ad valorem tax ("the CSBE Handbook" or AH566).[136] The Handbook, most recently updated as of January, 2015, is a petroleum valuation text written by knowledgeable oil & gas professionals and includes discussion of petroleum geology, the occurrence of oil & gas, the means and methods of production, and the procedures for valuing individual oil & gas properties. As part of that presentation, Appendix C to the Handbook is an extensive discussion of the derivation of discount rates, calculation of WACC discount rates, and the application of those discount rates to property valuation.

Both these publications are important and relevant for several reasons but primarily because they directly address the evaluation of individual oil & gas properties and bring the financially based WACC into the derivation of property discount rates. Together, the Texas Manual and the CSBE Handbook are consistent in their treatment of the issues and provide the means of applying the financial-based BFIT WACC to oil property valuation. Both are publicly available at no cost.

The derivation of a BFIT WACC discount rate is a comparatively straightforward exercise that, nonetheless, benefits from a working knowledge of the sources of the component parts and some experience in applying the required principles. The WACC has three major components combined in a relatively simple equation:

- Capital Structure – The proportion of debt (% Debt) and equity (% Equity) in total corporate capital.
- Cost of Debt ("COD") – The rate-of-return expected by the holders of the debt portion of total capital.
- Cost of Equity ("COE") – The rate-of-return expected by the holders of the equity portion of total capital.

$$\text{WACC} = [(\text{COD x \%Debt}) + (\text{COE x \%Equity})]$$

# Capital Structure

The Capital Structure is the combination of debt and equity used by the corporation to finance on-going operations including capital investments. Note that total capital is referred to as the basis of the capital structure and of the respective costs of debt and equity, and a public corporate source of data is implied. However, conclusions derived from public corporate sources should apply just as well to private companies, partnerships, and individuals; all must compete in the same markets for both funds and investment returns.

In the capital structure, **Debt** includes the outstanding amount of public and private debt plus capitalized leases as of or about the date of the analysis. WACC analysis usually includes only long-term debt that has maturity in excess of one year. Commercial paper and other short-term obligations are not included. A decision must be made whether to include the short-term (current year) portion of long-term debt.

The **Equity** portion of the Capital Structure is the market value of all outstanding common stock shares as of the date of analysis, not the balance sheet book value. Where a company has preferred stock issued, it may be treated as equity; however, some analysts consider preferred stock to be more like debt. The debate is academically interesting, but the amount of preferred stock issued by the oil industry is minuscule and has no measurable impact on the amount of either debt or equity.

The typical capital structure for oil companies can be shown to consist of about 30% debt and 70% equity based on market value. The proportions may vary with time as a function of changing tax laws, corporate policy and/or requirements, and interest rates. The maintenance of a balance between debt and equity is an important part of financial management in companies of all sizes.

As an example, at year-end 2020, the capital structure of a sample group of 19 publicly traded oil & gas companies had an average 43.52% debt and 56.39% equity.[137]

The capital structure that is appropriate for a COC discount rate is the proportion of debt and equity most likely to be used for future investments. Since most significant transactions tend to be cash or cash equivalents, the capital structure should reflect the mix of debt and equity maintained as corporate funds that is available for use in making investments.

## Cost-of-Debt

The Cost-of-Debt (COD) component consists of public and private instruments issued by the corporation, such as bonds and notes, and loans from banks, insurance companies and other institutions. The COD is the return demanded by the holders of the company's bonds and the interest rates on outstanding loans. Long-term debt is used because the term of the debt is comparable to the term of an investment in oil & gas properties. The yield-to-maturity (YTM) on debt is used because YTM is a forward rate that continuously measures the expectations of investors regarding the debt of specific companies.

The overall COD for a company or a group of companies is calculated as an average of the interest rates and (YTM) on each debt instrument, weighted by the dollar amount outstanding. A list of the outstanding debt issues can usually be obtained from the company annual report and/or SEC 10-K (sometimes supplemented with other data) where the outstanding amounts for each debt issue are tabulated. The YTM for public issues (bonds, debentures) can be obtained from the monthly Moody's Bond report or similar publication. The effective interest rates on institutional debt can be taken from the annual report or is calculated if sufficient information is accessible. The COD for the company is the weighted average of the yield and interest rates for the individual issues as a function of total debt.

## Cost-of-Equity

The Cost-of-Equity (COE) is commonly defined as the market value of the outstanding common shares of the company on the date of evaluation. **Retained Earnings** are also equity since they are the property of the common shareholders. Retained earnings of the corporation are invested by company management on behalf of the shareholders for the purpose of increasing shareholder wealth. As might be expected, it is not easy to determine the return desired by equity holders. Even the definition is difficult.

Weston says:[138] "*the cost of equity capital is defined as the rate of return that must be earned to prevent the price of the stock from falling....*" Mullins states, "*The finance literature defines the cost of equity as the expected return on a company stock. The stock's expected return is the shareholder's opportunity cost of the equity funds employed by the company."*[139] and, *"In theory, the company must earn this cost on the equity-financed portion of its investments or its stock price will fall. If the company does not expect to earn at least the cost of equity, it should return the funds to the shareholders who can earn this return on other securities at the same risk level in the financial marketplace."*[140] Other authors provide similar definitions.

COE is, therefore, an expectation rate set by investors. If the company return-on-equity falls below expectations, investors will sell their shares, thereby depressing the price of the stock, which reduces market value, which results in diminished corporate net worth. In order to avoid this chain of events, companies attempt to invest funds in projects that are expected to earn a return at least equal to the WACC. In practice, minimum required returns may be set above the WACC in order to compensate for those projects that fail to provide the anticipated returns. Financial texts describe four basic methods for estimating a forward COE:

- Earnings/Price Ratio
- Capital Gains Valuation Method (also known as Realized Investor Yield)
- Dividend Growth Model
- Capital Asset Pricing Model

The first three methods are perfectly sound approaches to COE and have been used in many COC studies, but there are two problems in using these methods in an industry-specific study. First, many companies do not issue dividends, and earnings relative to price may fluctuate so much that they are less than useful for projection of future earnings or price. Second, in order to be useful in estimating anticipated returns, projections of future earnings, prices, and dividends are necessary. Unless one is a full-time equity analyst, such projections are not usually possible. Third-party projections can often be obtained from tested and reliable sources, but not for all companies.[141]

The Capital Asset Pricing Model (CAPM) has become the primary method of estimating the cost of equity. Originally developed and published in the 1960s, the model was a simple (straight line) mathematical construction. However, over time the model has been revised, expanded, embellished, and debated so that there are now numerous versions that attempt to include many different factors that are thought to influence the COE. Examination of even the simplest original version of CAPM is beyond the scope of this discussion, but both the Texas Manual and the CSBE Handbook include extensive discussion of CAPM.

## Estimating the Corporate Weighted Average Cost of Capital

Given sufficient information from reliable sources, calculation of an industry-wide cost-of-capital is tedious but not onerous.

As an example, assume that at year-end 2020, a sample group of 19 publicly traded oil & gas companies has a capital structure of 43.52% debt and 56.39% equity; the cost of debt was determined to be 4.20% while the cost of equity is 18.36%. Using this data, the WACC is calculated as follows:

BFIT WACC = (%Debt X Cost of Debt) + (% Equity X Cost of Equity)
= (0.4352 X 4.20) + (0.56.39 X 18.36)
= 12.45%

The above calculation is taken from data included in the *2021 School District Property Value Study: Discount Rate Range for Oil and Gas Properties* dated November, 2021.[142] The Property Value Study (PVS) is published annually, as required by state law, by the Property Tax Assistance Division (PTAD) of the Comptroller of Public Accounts as an adjunct to the Texas Manual. The PVS has been conducted annually since the 1980s and is the only known publicly available study of WACC discount rates for application to oil & gas property evaluation. The WACC is calculated from year-end financial data for a sample group of publicly traded petroleum companies. The results of the studies for year-end 2013 through year-end 2020 are tabulated below. These results provide a solid foundation for derivation of property valuation discount rates.[143]

| Year Ending | Sample Group | Long Term Debt % of Capital | Equity % of Capital | Cost of Debt | BFIT Cost of Equity | BFIT WACC |
|---|---|---|---|---|---|---|
| 2020 | 19 | 43.52 | 56.39 | 4.2 | 18.36 | 12.45 |
| 2019 | 18 | 30.44 | 69.55 | 4.14 | 15.2 | 11.82 |
| 2018 | 18 | 27.84 | 72.15 | 5.12 | 15.76 | 12.75 |
| 2017 | 17 | 20.69 | 79.3 | 4.06 | 14.99 | 12.62 |
| 2016 | 18 | 20.87 | 79.12 | 4.5 | 17.37 | 14.64 |
| 2015 | 18 | 30.28 | 69.64 | 6.4 | 16.91 | 13.94 |
| 2014 | 18 | 21.56 | 78.4 | 4.28 | 16.01 | 13.46 |
| 2013 | 18 | 18.35 | 81.62 | 4.48 | 17.52 | 15.09 |

**Table 9.** Results of the Property Value Study (PVS)

## Adjustments to the BFIT WACC Discount Rate

The BFIT WACC can be used as the baseline from which to build a property- or project-specific discount rate. However, discount rates derived from WACC analysis cannot be applied directly to evaluation of individual oil properties. The WACC is deficient in several elements which must be considered before application can be made to a specific property. There must first be adjustments to account for inherent differences between WACC returns and the returns that would be expected from a specific property. The differences are several, complex, and important.

1. Oil & gas production is a wasting asset; properties are produced to depletion and/or economic limit. Petroleum companies are not considered to be wasting assets; companies expand, add product lines, and develop new properties. Recovery or Reversion of capital invested in corporate bonds or common stock can be obtained by real-time sale in the respective markets. By contrast, in producing oil properties, reversion of capital invested can only obtained from the cash flow from the property.
2. The return derived from WACC analysis represents the return from a portfolio of debt and equity issued by public companies where reversion is presumed but not provided. Data from private companies would be expected to indicate competitive but higher returns.
3. The WACC is the discount rate used to value the company, not the individual assets of the company. The WACC may well be used as a benchmark or minimum rate against which to compare investments, but each investment has risk that may not be the same as the composite risk of the firm.
4. Individual oil property investments carry more risk than an investment in a corporation. An oil property is an investment dependent on one income stream derived from oil & gas production from that property. On the other hand, the components of WACC return represent returns from a wide portfolio of investments so that risk of loss due to failure of one asset is minimized by the income available from all other assets.

The return anticipated from a specific producing property must exceed the WACC to account for deficient areas which can include, but are not limited to, Liquidity, Return-of-Capital, and Portfolio Diversity.

## Liquidity

WACC is derived from the stocks and bonds of publicly traded companies. The COD and COE are the returns that are expected by investors in publicly traded debt and equity. These investments are highly liquid; they can be bought and sold almost instantaneously. The same cannot be said of individual oil properties. This illiquidity demands a premium on the cost-of-capital if it is to be used to value properties.

## Return-of-Investment

Investors generally require a return-of their investment and a return-on that investment (interest). The WACC, as commonly constructed, represents only a return-on-capital invested. The calculation of a WACC, at a specific time, assumes that the calculated return will continue from that time forward – in effect, a level and perpetual annuity (return-on-investment). The WACC assumes that the original investment is recovered when the debt or equity interest is sold. In finance, this is termed **return-of-capital**.

In contrast, individual oil & gas properties have variable and declining income streams and are produced to depletion. There is a fixed volume of oil & gas; only a portion of the volume can be produced and sold. The production of part of the volume reduces the amount remaining to be produced. When all economically producible oil has been recovered, the asset will have been diminished to zero and will have no residual or reversionary value. This will be true even though changes in economic and technological conditions may act to defer the point of final depletion. There is no remaining asset value from which to recover the original investment. The income stream must provide both a return-on-investment and a return-of-investment, the combination of which must necessarily be greater than the WACC return alone.

The requirement for return-of-investment from the income stream is demonstrated by the use of the payout method as a common criterion for oil & gas investment decision making. In the Payout approach, the time required for recovery of invested capital from the income stream is calculated and then used to judge the acceptability of an investment. The time period anticipated for Payout varies but historically averages 3-5 years. All income after payout is considered a return above payout and can be equated to a return-on-investment.

This circumstance was recognized long ago by real estate appraisers. While a couple of methods to account for return-of-investment from cash flow are recognized, the Hoskold Method is most useful. The method was developed to aid in appraisal of coal mines and has direct application to oil & gas appraisal. Hoskold assumes that funds are set aside from cash flow and retained in an account earning a safe rate of interest.

## Portfolio Diversity

The largest difference between corporate cost-of-capital and the returns from investments in individual oil properties is the diversification of risk through a "Portfolio Effect." A company that owns a number of properties can expect an income stream from each, whereas a single property provides only one income stream. In the corporate context, there is greater possibility of loss in the single income stream than in the multiple streams where a failure of one stream has a limited effect and may be offset by an increase in another income stream. The income stream from a particular property may have greater or lesser risk than the other income streams in the portfolio as the result of any number of engineering, geologic, operational, or economic factors.

## Example

Some years ago, while working on a property tax appeal for a major oil company (name withheld to protect the innocent), the subject of portfolio diversity and the discount rate came up. An executive for the company noted, and I paraphrase, *"Did you know that out of all the stuff we sell, gasoline, oil, whatever, our most profitable income stream is from Twinkies in the convenience stores."* This may be true, more probably apocryphal, (some experts claim that hot dogs are the most profitable), but it illustrates the point that the discount rate derived from public company WACC is based on income from a great many sources, while the return derived from an oil production property has only one source of income.

# Adjusting the Cost-of-Capital Discount Rate

An investor expects a return commensurate with the risk taken in the investment. The WACC can serve as a base rate for discounting but must be increased to account for the additional risk inherent in investment in a single oil property. The question is how to make a reasonable adjustment to the WACC discount rate that captures some or all (not likely) of the Liquidity, Return-of-Investment, and Portfolio Diversity without having to delve back into arcane financial discussions. While there are many scholarly papers in finance and other fields that discuss these issues, they rarely apply directly to oil & gas evaluation. It is difficult to separate the discount rate premium into discrete factors for Liquidity, Return-of-Investment, and Portfolio Diversity. The empirical data that does exist is too limited in both breadth and depth to allow detailed dissection; however, the knowledge that the differences exist is a sufficient starting point for further analysis.

Studies of data from California petroleum property transactions over the period of 1985 through 2005 consistently show that the discount rate derived from the transactions exceeded the BFIT WACC for a sample group of oil & gas companies by an average of 7.05%. The transaction data was collected for over 150 acquisitions of properties with apparently low-risk production over the 20-year period; the BFIT WACC discount rates were derived each year in accordance with the CSBE Handbook and essentially matched the Texas PTD data for the same year. While the composition of the difference can be (and has been) debated, the fact that the difference exists argues for the need to adjust the WACC discount rate.

Surveys of oil & gas companies, and a review of acquisition evaluations, suggest that Payout remains a useful, if subordinate, investment criteria. The payout approach simply determines the time required to recover the original investment from the anticipated cash flow of the project or property. Payout provides for the return-of-investment by estimating the time necessary to accumulate the original investment from cash flow. All income received after Payout contributes to return on investment. An acceptable Payout period might be 3 to 5 years. If the payout period was 5 years, then one could infer an average annual return-of-investment of 20% for those five years. In real estate evaluation, the concept of Recapture Rate is well known and contemplates the use of a fixed or variable rate as part of a built-up or total discount rate.

## Pure-Play Analysis

Another approach to evaluating the difference between the WACC discount rate and the empirical data from market sales is to consider discount rates derived from so-called Pure-Play Analysis, a methodology commonly used in business valuation. The theory is that the discount rate for producing properties can be approximated by determining the cost-of-capital for companies whose business income is entirely, or in large part, from oil & gas production: no refineries, pipelines, or AM/PM stores. If a sample group of publicly traded companies whose business consists solely of oil production is assembled, the WACC for those companies could be used as a proxy for a specific property discount rate.

In 1997, Ibbotson Associates completed a study for the Western States Petroleum Association (WSPA) entitled, "*WACC for Pure-Play Oil and Gas Extraction and Refining Entities*."[144] In that study, Ibbotson estimated a cost-of-capital for two groups of companies whose business was 100% production and, separately, 100% refining. The COC was calculated using a standard WACC approach, where the COE was derived using adjusted CAPM. As applied in the evaluation of petroleum properties, the discount rate is composed of a return-of-investment and a return-on-investment. The return-of-investment is the repayment or recovery of the initial investment. The return-on-investment, "*combines a safe rate* [of interest] *with a premium to compensate the investor for risk, the burden of management, and the illiquidity of invested capital*."[145]

The COE was then combined with the cost-of-debt and a typical structure to calculate an AFIT weighted average cost-of-capital of 9.51% for a hypothetical pure-play producing company, and 10.87% for a pure-play refining company. After adjusting for liquidity using standard business valuation methods and for income tax, the Ibbotson analysis obtained a Before Tax WACC of 20.15% for pure-play producing companies, and 23.04% for pure-play refining companies. These results suggest that, when liquidity is taken into account, a company that receives all of its business income from oil production has an effective discount rate of 20.15% BFIT. The Pure-Play analysis did not attempt to account for portfolio diversity.

# APPENDIX D
# CONSIDERATION OF RISK AND UNCERTAINTY IN PETROLEUM PROPERTY EVALUATIONS

There is a degree of ambiguity about including a discussion of risk and/or uncertainty in a treatise on property evaluation. The ambiguity is not whether to delve into the subjects but rather the depth and detail to be allotted. There is no question that any oil property evaluation involves some concern that the data being used, the analysis of that data, and the application of the data to estimate a future income stream and estimate the economic value for a property or project may be subject to some degree of uncertainty. Several industry authors have written extensively about the relation of risk and uncertainty to oil property evaluation. A full discussion of the subject, including Probability Theory and the various techniques for application of probability and other issues to the evaluation of risk and uncertainty in petroleum properties, is presented by those authors. It is notable, however, that most of the authors and their publications pertain primarily to exploration. A leading text on the subject which will be referenced hereafter is *"Decision Analysis for Petroleum Exploration"* by Newendorp and Schuyler (2000),[146] but it is only one in a series of examinations of the subject starting with Grayson[147] and Kaufman[148] through Megill[149] and Rose[150] to the present day. In addition, oil & gas exploration has proven to be a useful laboratory for those studying the use of statistics and probability analysis applications to real world decision making (Macmillan).[151]

It is not the intent of this section to attempt to duplicate any of those works in whole or in part; the subject is extensive and beyond the scope of this volume. As with the geology of petroleum and the workings of reservoir engineering, a brief summary coupled with references should be sufficient.

This a good point to consider whether risk and uncertainty are the same thing. Newendorp offers the following: *"Most people consider the words "risk" and "uncertainty" to be synonymous or nearly so. There is widespread disagreement, and attempting precise definitions will invite arguments. Despite this apprehension, I will share "loose" definitions to convey the manner in which I most often use these words."*

*"Risk: The potential for outcomes that are quite different. Typically, this applies to a discrete event, such as the result of a test well; Discovery or Dry Hole! Most people think of risk as downside potential. In recent years, people, especially in project risk management, have sought to differentiate the potential either good or*

*bad impacts. Thus, risk refers to a contingency that can be either bad ("threat") or good ("opportunity")."*

*"Uncertainty: The characteristic of a value is variable. Typically, I apply this to parameters where the result can be realized over a range of possibilities. For example, oil price, where we know there is going to be an oil price, and the uncertainty is 'how much.'"* In regard to the latter example, not just how much initially but how could oil price change over the life of the project.

Newendorp further notes, *"Uncertainties about future events (crude sales price fluctuations, future tax and government policies, storm damage, etc.) also contribute to the overall risk.... These uncertainties taken in combination imply that there can be more than one outcome to any decision alternative.... Decision making under uncertainty always involves at least two possible outcomes for at least one decision alternative.... By comparison, if we knew exact values for all the parameters we would be able to calculate an exact value of project profitability. Such a calculation would be called a "deterministic" value of profitability.... If the consequences of all possible decision alternatives could be calculated exactly the process of decision making would be much simpler. It is the unknown resulting from our inability to measure or predict values... that makes the decision-making process complicated ... and interesting."*

The definitions offered by Newendorp et al. are more in the order of "working criteria" that may be useful in discerning a difference if one is needed. Drawing on those sources, **risk** might be defined as applying to the potential or possibility that an event could have *"outcomes that are quite different"* and, possibly, mutually exclusive. This suggests an outcome that is either X or Y. The most obvious example is drilling an exploratory well – it is either successful or not – depending on the prevailing definition of successful. Risk could be interpreted to refer to a contingency where the outcome is either "good" or "bad." Success may depend on expectations; the well was dry (expectations) but it provided useful information.

**Uncertainty** is less stark. Uncertainty applies to events that are themselves reasonably predictable to occur but where the outcomes are not necessarily predictable. The examples of uncertainty in oil property evaluation are as numerous as there are variables in the evaluation. Some of these variables, such as production rates and product prices, have been discussed previously in the context of uncertainty. Probably (there is that word again) the best way to separate risk from uncertainty is to say that Risk is the likelihood that something will occur but the result is unknown. Conversely, uncertainty or confidence refers to something that is expected to occur but there is a range of possible outcomes which are not equally likely to occur. Example: Some evaluators refer to price risk regarding future oil prices, but there is no risk as such; prices will occur, but there is uncertainty about how much it might vary over time.

The discussion of oil property evaluation has made no distinction regarding the type of property or stage of development of the property/project, but there has been an underlying implication that the property is developed and producing; we are not dealing with exploration projects. In the context of oil property economic evaluation, uncertainty is the more applicable condition than is risk.

Within any property evaluation there are numerous opportunities for uncertainty to impact the income stream and economic benefit attributable to the property. The estimation of potential production and the projection of that production can be done using all available data and the best methods, but the actual production is still subject to the vagaries of a natural system – the reservoir rocks and fluids – that cannot even be sampled and measured to any major extent, let alone be accurately modeled to eliminate the potential for variance from expectations.

The evaluation of an existing property with a known performance history, geology, production mechanism, and other parameters, could be done with a reasonably high level of confidence that included allowance for uncertainty in production rates, etc. On the other hand, the application of new production processes (hydraulic fracturing w/ horizontal wells, enhanced recovery, in-fill drilling) to that same property would introduce an entirely new level of uncertainty to production projections, operating costs, and investment schedules, among others.

Having said that, it should also be noted that the large majority of evaluations will recognize that uncertainty, maybe even a little risk, is inherent in the evaluation, but in most cases, there will be no definable effort to account for the uncertainty. Rather, the evaluation is done using a "best estimate" approach using available data and applying basic analysis and experience to the selection of production rates, reservoir performance, operating costs, product prices, etc. The result may be one or two estimates of production, income streams, and, if needed, value. This is the **deterministic approach** to evaluation. It is an approach that serves the purpose well and frankly accounts for the large majority of economic evaluations done. It is particularly useful for evaluation of ongoing operations on existing properties. The emphasis in this approach is the careful use of available data and particularly on experience. The approach does not ignore uncertainty but absorbs it into the process.

However, there are often evaluations that require the application of processes that explicitly consider the degree of uncertainty in the data and/or the interpretation of that data. While a deterministic approach may work satisfactorily for an existing group of wells on a property where there is long-established production and operations, the approach is probably not acceptable for evaluation of that same property where expanded drilling and EOR are planned. In the latter case, the implementation will require considerable investment, changes in operations, and projection of development and production significantly different from current operations.

The most commonly accepted method for evaluation of a project where explicit consideration of risk and uncertainty is necessary, or at least desirable, is **expected**

**value theory** or one of the many variants. In the simplest form expected value is the probability-weighted value of all possible outcomes (Newendorp). For situations with several possible outcomes, the expected value is the sum of each outcome value times the probability of occurring. A simplistic example: If a well can be evaluated to produce two different income streams (using the same data) but there is some uncertainty as to which outcome could occur, an expected value estimate could be as follows:

Outcome A = $1,000,000 with a 35% chance to occur.

Outcome B = $800,000 with a 65% chance to occur.

The expected value of the exercise would be ($1,000,000) (0.35) + ($800,000) (0.65) = $870,000

Depending on the stage of development and history of production of the property being evaluated, along with the time and resources available, a property valuation should attempt to recognize and account for uncertainty. There are several methods of analyzing uncertainty and applying adjustments. The process has steps which can be generalized:

1. Define the source of the uncertainty.
2. Determine if the uncertainty is measurable.
3. Define a range of values for the uncertainty.
4. Select an evaluation method(s).
5. Apply appropriate adjustments to the evaluation.

# REFERENCES


1. Disputes on this issue are the source of many lawsuits.

2. *Manual of Oil and Gas Terms*, 13th Edition; Martin, Patrick H. and Kramer, Bruce M., 2006

3. "Interests in property vary, so real property is said to include a 'bundle of rights' inherent in the ownership of real estate. Ownership rights include the right to use real estate, to sell it, to lease it, to enter it, to give it away, or to choose to exercise all or none of these rights." *The Appraisal of Real Estate*, 11th Edition, 1996, Appraisal Institute, p. 7

4. Paraphrased from "Manual of Oil & Gas Terms," Martin, Patrick H., and Kramer, Bruce M., 2006, p. 1109

5. "Manual of Oil & Gas Terms," Martin, Patrick H., and Kramer, Bruce M., 2006, p. 780

6. Will-Drill Res. V Huggs, 738 So. 2d 1196 (La. Ct. App. 1999)

7. *The Appraisal of Real Estate*, 11th Edition, Appraisal Institute, Chicago, IL, p. 90

8. "Decline Curve Analysis Using Type Curves," Fetkovich, M. J., SPE Journal of Petroleum Technology, June 1980, page 1065

9. "Decline Curve Analysis Using Type Curves," Fetkovich, M. J., Vienot, M. E., Bradley, M. D., and Kiesow, U. G., SPE 13169, September 1984, Dallas, TX

10. "The TM Sand Stimulation Project," Yoelin, Sherwin D., SPE 3194-PA, August 1971

11. "Mechanics of Secondary Oil Recovery," Smith, Charles Robert, Reinhold Publishing Corporation, New York, 1966 and "Steamflood Reservoir Management," Hong, K. C., Pennwell Books, Tulsa, OK 1994

12. The graph was prepared by InflationData.com as of November 10, 2022.

13. There is another form known as a "forward contract" which acts like a spot contract but may be an agreement for purchase at a future date. The price may be specified, but the volume and delivery date may vary.

14. Defined by EIA as *"Specific domestic crude with 0.42% sulfur by weight or less, not less than 37° API gravity not more than 42° API."*

15. Futures contracts can be used for most commodities. For instance, it is not uncommon for airlines to make use of futures contracts to manage jet fuel requirements.

16. *Maximizing Field Value with a Royalty Rate that Tracks Oil Price,* Mercier, David; Mount, Paul, II; Willard, Al; SPE 37951, Journal of Petroleum Technology, Vol. 49, pp. 1220-1230, November 1997

17. Council of Petroleum Accounting Societies, Wichita, KS

18. American Petroleum Institute, 200 Massachusetts Ave., Washington, D.C.

19. As an example: *"Draft Report: Status of Petroleum Production Industry in California: The Cost of Regulatory Compliance,"* prepared for the California Department of Conservation by Foster Economics et al., San Francisco, CA, 1993. No final report was issued.

20. At the risk of adding a new acronym to an already overburdened lexicon, AR&R is used in this discussion for three reasons. First, it is the terminology that has been in use in California and has survived litigation and regulatory interpretation for about 40 years.

Second, the terms allow separation of the process into three logical segments related to the sequence of events relative to regulatory procedures. Third, it allows the costs of compliance with regulations to be estimated and applied at the appropriate time in the evaluation.

21. *Department of Conservation Idle Well Program Report,* January 1, 2018 – December 2018, California Department of Conservation, Sacramento, CA

22. Ibid

23. Ibid

24. Ibid

25. Statement of Financial Accounting Standards (SFAS) as amended (http://FASB.org/pdf/aop_Fas143.pdf)

26. *"The Appraisal of Real Estate,"* 11th Edition, Appraisal Institute, Chicago, IL, p. 496

27. *Well Abandonment in the Los Angeles Basin: A Primer,* Evans, B.L., Evans, Frey & Walker: Sailor, R. P., R.P. Sailor & Associates; Santiago, E., California Division of Oil, Gas, and Geothermal Resources; SPE 83443, May 2003, Dallas, TX

28. Two studies are informative if a bit biased: "Waiting for Crisis: Regulatory Inaction and Ineptitude and the Guadalupe Dunes Oil Spill," Beamish, Thomas D., Social Problems, Vol. 49, No. 2, pp. 150–177 and "Guadalupe oil field remediation and abandonment project," Arthur D. Little Consulting for the Lead Agency: Department of Planning and Building, San Luis Obispo County, CA

29. Operators in areas where remediation may be an issue will usually have conducted at least preliminary assessments known under California Environmental Quality Act as Phase I and Phase II assessments. These studies may indicate the level and extent of concern. A Phase III assessment goes farther to quantify issues and includes cost estimates.

30. *"Measuring the Effects of Hazardous Materials Contamination on Real Estate Values: Techniques and Applications,"* a Technical Report of the Appraisal Institute, Chicago, IL, 1992

31. *"Environmental Site Assessments and Their Impact on Property Value: The Evaluators Role,"* Colangelo, Robert V. and Miller, Ronald D., Appraisal Institute, Chicago, IL, 1995

32. "Assessors' Handbook 566 – Assessment of Petroleum Properties," California State Board of Equalization, Sacramento, August 1996, pp. 6–8

33. *Survey of Parameters Used in Property Evaluation,* Society of Petroleum Evaluation Engineers, June 2022, Houston, TX

34. *"Investment Analysis: U.S. Oil and Gas Producers Score High in University Survey,"* Boyle, Hugh F., and Schenck, George K., "Journal of Petroleum Technology," Vol 37, April 1985, p. 680

35. *"Current Investment Practices and Procedures: Results of a Survey of U.S. Oil and Gas Producers and Petroleum Consultants,"* SPE #25824, Dougherty, E. L. and Sarkar, Jayati, University of Southern California, presented at Hydrocarbon Economics and Evaluation Symposium, March 1993, Dallas, TX

36. *Cost of Capital: Estimation and Application,* Pratt, Shannon P., John Wiley & Sons, Inc., 1998, p. 3

37. *Principles of Corporate Finance,* Brealey, Richard A. and Myers, Stewart C., 5th Edition, McGraw-Hill, 1996, p. 206

38. *"Principles of Corporate Finance,"* Brealey, Richard A., Allen, Franklin, and Myers, Stewart C., 2nd Edition, 2011, McGraw Hill, New York

39. *"Managerial Finance,"* Weston, J. Fred and Copeland, Thomas E., 8th Edition, 1986, Dryden Press, New York, p. 99

40. *"Financial Management for Decision Making,"* Bierman, Harold and Smidt, Seymour, 1986, MacMillan Publishing Company, New York, p. 139

41. *"Principles of Corporate Finance,"* Brealey, Richard A., Allen, Franklin, and Myers, Stewart C., 2nd Edition, 2011, McGraw Hill, New York, p. 23

42. *"Cost of Capital: Estimation and Application,"* Pratt, Shannon P., 1998, John Wiley & Sons, Inc., New York, New York

43. *"Principles of Corporate Finance,"* Brealey, Richard A., Allen, Franklin, and Myers, Stewart C., 2nd Edition, 2011, McGraw Hill, New York, p. 23

44. *"Managerial and Engineering Economy,"* Taylor, George A., D. Van Nostrand Company, Inc., 1964, p. 13.

45. *The Appraisal of Real Estate,* 11th Edition, The Appraisal Institute, Chicago, IL, 1996, p. 44

46. *Principles of Corporate Finance,* 2nd Edition, Brealey, Richard A., Myers, Stewart C., and Allen, Franklin; McGraw-Hill Irwin, 2011, p. 23

47. *Fundamentals of Corporate Finance,* 6th Edition, Brealey, Richard A., Myers, Stewart C., and Allen, Franklin; McGraw-Hill Irwin, 2009, p. 379

48. *"Cost of Capital: Estimation and Applications,"* Pratt, Shannon P., John Wiley and Sons, Inc., 1998, p. 5

49. Ibid, p. 5

50. *"Analysis and Management of Petroleum Investments: Risk, Taxes, and Time,"* Campbell, John M., Jr., 1987, John M. Campbell and Co., p. 67

51. *2021 School District Property Value Study, Discount Rate Range for Oil and Gas Properties,* Texas Comptroller of Public Accounts, Property Tax Assistance Division, Austin, TX, November 2021, p. 3

52. *"Selection of a Discount Rate or Minimum Acceptable IRR,"* SPE 16843, Sinha, M.K. and Poole, A.V., Morgan Guarantee Trust, September 1987

53. *"Thirty-Third Annual Survey of Parameters Used in Property Evaluation,"* Society of Petroleum Evaluation Engineers, June 2014, Houston, TX, p. 32. The 2014 report is used because of the clarity of text; previous and subsequent reports provide similar information.

54. Ibid, p. 32

55. Ibid, p. 36

56. Ibid, p. 36

57. Ibid, p. 36 The Potential Risk Factors are not defined in the survey questionnaire or the report.

58. *2021 School District Property Value Study, Discount Rate Range for Oil and Gas Properties,* Texas Comptroller of Public Accounts, Property Tax Assistance Division, Austin, TX, November 2021

59. Some experts have referred to this as the testosterone effect often epitomized by the determination to acquire a specific property regardless of the cost, commonly seen in bidding contests for properties or companies.

60. *Investment Analysis: U.S. Oil and Gas Producers Score High in University Study*, Boyle, Hugh F., Jr. (Manufacturers Hanover Trust Co.) and Schenck, George K. (Pennsylvania State University), January 1984; and SPE Journal of Petroleum Technology, April 1985, p. 680–690

61. *Current Investment Practices and Procedures: Results of a Survey of U.S. Oil and Gas Producers and Petroleum Consultants*, E. L. Dougherty and Jayati Sarkar, presented at SPE Hydrocarbon Economics and Evaluation Symposium, March 29-30, 1994; and published as SPE 25824 by Society of Petroleum Engineers, Dallas, TX

62. *"Thirty-Fifth Annual Survey of Parameters Used in Property Evaluation,"* Society of Petroleum Evaluation Engineers, June 1, 2016, Houston, TX, p. 28. The 2016 report is used because of the large number of responses (190) as compared to earlier surveys and clarity of text; previous and subsequent reports provide similar information in various formats.

63. *"Thirty-Third Annual Survey of Parameters Used in Property Evaluation,"* Society of Petroleum Evaluation Engineers, June 1, 2014, Houston, TX, p. 25

64. "Fundamentals of Corporate Finance," Brealey, Richard A., Myers, Stewart C., and Marcus, Alan J., 6th Edition, 2009, McGraw-Hill Irwin, New York, NY

65. Ibid, p. 223

66. Ibid, p. 225

67. Ibid, pp. 232–233

68. Ibid, p. 230

69. The Profitability Index or Ratio is referenced by some texts and sources as the Return-on-Investment (ROI) and can be confused with the Internal Rate-of-Return. The two are not compatible.

70. BM&M, P. 245

71. BM&M, P. 247; Original data reprinted from Journal of Financial Economics, Vol. 60, Issue 2-3, Graham, J. R. And Harvey, C. R., "The Theory and Practice of Corporate Finance: Evidence from the Field," May 2001, pp. 187–243

72. *Manual for Discounting Oil and Gas Income*, Texas Comptroller of Public Accounts, June 2021, Austin, TX

73. Lynch v. State Board of Equalization, 164 Cal. App. 3rd 94 (Cal. Ct. App. 1985)

74. *Assessors' Handbook 566 – Assessment of Petroleum Properties*, California State Board of Equalization, Sacramento, CA, August 1996

75. Estate Tax Regulation 20.2031-1(b) and Gift Tax Regulation 25.2512-1

76. Posgate, Louis R., An *Oil and Gas Federal Tax Case Review: Fair Market Values with Volatility*. Business Valuation Review, American Society of Evaluators, December 2002, p. 178

77. *"The Appraisal of Real Estate,"* 11th Edition, 1996, The Appraisal Institute, Chicago, IL, p. 24

78. Ibid, p. 26

79. Ibid, p. 26

80. Ibid, p. 26

81. Ibid, p. 26

82. Babcock, Henry A., Appraisal Principles and Procedures, Richard D. Irwin, Inc. 1968

83. *The Appraisal of Real Estate,* 11th Edition, 1996, The Appraisal Institute, Chicago, IL, p. 20

84. Ibid, p. 21

85. Ibid, p. 21

86. Ibid, p. 22

87. United States v. Miller, 317 U.S. 369, 374 (1943) and echoed by Mr. Justice Cattanach, Federal Court (Trial Division), Canada, in Henderson Estate and Bank of New York v. M.N.R., 73 DTC 5471, p. 5476

88. Black's Law Dictionary, 2nd Pocket Edition, 2016

89. State v. Carpenter, 89 S.W. 2d 194 (Tex.1936)

90. West Texas Hotel Co. vs. City of El Paso, et al., 83 S.W. 3d 772 (Tex. Civ. App. – El Paso 1935), writ dismissed.

91. Howell, Jay D., Jr., A Texas *Practice,* Vol. 21, Property Taxes, 4th Edition, West Group

92. See Howell, pp. 387–388 in re City of Austin v. Cannizo, 153 Tex 324, 267 S.W. 2d 808 (1954)

93. *Uniform Appraisal Standards for Federal Land Acquisitions,* Section 4.2.1 – Definition of Market Value, Interagency Land Acquisition Conference, The Appraisal Institute, Chicago, IL, 2016, p. 93

94. United States v. Reynolds, 397 U.S. 14, 16 (1970)

95. United States v. Miller, 317 U.S. 369, 374 (1943)

96. United States v. Cartwright, 411 U.S. 546, (1973) citing Treasury Regulation 20.2031-1(b)

97. Canada Revenue Agency, *Fair Market Value,* Policy Statement, CSP-F02, October 25, 2002

98. *Canada, Defining Value: What is Fair Market Value?* Neil de Gray, Duff & Phelps, September 26, 2013.

99. California Revenue and Taxation Code,110(a)

100. Rule 2 cited in Assessors' Handbook Section 501, Basic Appraisal, California State Board of Equalization, January 2002, p. 10

101. Assessors' Handbook Section 501, Basic Appraisal, California State Board of Equalization, January 2002, p. 10.

102. California Code of Civil Procedure, Eminent Domain, C.C.P, 1263.320(a)

103. *Texas Property Tax Code,* General Provisions, Section 1.04August 2002)

104. *The Appraisal of Real Estate Appraisal,* 11nd Edition, The Appraisal Institute, Chicago, IL, 1996, p. 22

105. ASA Business Valuation Standards, Definitions, American Society of Appraisers, November 2009, p. 27

106. Federal Register, vol. 55, no. 163, August 22, 1990, pp. 34228 and 34229 and *The Appraisal of Real Estate Appraisal,* 11nd Edition, The Appraisal Institute, Chicago, IL, 1996, p. 23

107. "*Valuation of Oil and Gas Reserves,* Arps, Jan J., in Petroleum Production Handbook, vol. II, Thomas C. Frick, Editor in Chief, McGraw-Hill, 1992, p. 38-5

108. *Which Fair-Market-Value Method Should You Use?* Garb, Forrest A., Journal of Petroleum Technology, January 1990, p. 8

109. Canadian Oil and Gas Evaluation Handbook, Vol. 1, *Reserves Definitions and Evaluation Practices and Procedures*, SPEE Calgary Chapter, June 2002, pp. 2-6

110. *Valuation of Oil and Gas Reserves by the Comparative Method*, William E. Strevig, SPE 18907, Dallas, TX, 1989

111. *Which Fair-Market-Value Method Should You Use?* Garb, Forrest A., Journal of Petroleum Technology, January 1990, p. 8

112. *An Examination of Fair Market Value Concepts for Oil and Gas Properties*; Williams, Joseph, The Appraisal Journal, July 1991

113. Williams, Joseph; Fair Market Value, Theory and Practice; SPE 22620, presented at SPE ATCE, October 6-9, 1991, Dallas TX

114. *An Examination of Fair Market Value Concepts for Oil and Gas Properties*; Williams, Joseph, The Appraisal Journal, July 1991, p. 402

115. Ibid. p. 402

116. Ibid. p. 404

117. Ibid. p. 410

118. F. J. Dingle and A. David, "The Perils of Market Value Averages," Proceedings of the 1987 SPE Annual Technology Conference and Exhibition (SPE paper 16841, 1987): 75-80

119. A. David and T. S. Hickman, "What is the Market Value of Long-Life Reserves?", Proceedings of the 1990 SPE Annual Technology Conference and Exhibition (SPE paper 205081, 1990): 377-385

120. *"Market Value: What Does it Really Mean?"*, Sanders, Michael A., The Appraisal Journal, Summer, 2018, p. 206

121. Byrl N. Boyce, compiler and editor, *Real Estate Appraisal Terminology*, (Chicago: American Institute of Real Estate Appraisers and Society of Real Estate Appraisers, 1975), p. 137

122. Byrl N. Boyce, compiler and editor, *Real Estate Appraisal Terminology*, (Chicago: American Institute of Real Estate Appraisers and Society of Real Estate Appraisers, 1981), pp. 160-161

123. American Institute of Real Estate Appraisers, The Appraisal of Real Estate, 9th Edition, Chicago, 1987, pp. 18-21

124. American Institute of Real Estate Appraisers, The Appraisal of Real Estate, 11th Edition, Chicago, 1996, pp. 21-22

125. County of Clark v. Buckwalter, 115 Nev. 58 (April 1999)

126. *Uniform Appraisal Standards for Federal Land Acquisitions*, Section 4.2.1 – Definition of Market Value, Interagency Land Acquisition Conference, The Appraisal Institute, Chicago, IL, 2016, p. 93

127. *"Market Value: What Does it Really Mean?"*, Sanders, Michael A., The Appraisal Journal, Summer, 2018, p. 216

128. Ibid, p. 216

129. Western States Petroleum Association, reports of *Fair Market Value Transactions, Cost of Capital, and Risk*, prepared annually 1983 through 2005, Glendale, CA

130. Texas Property Tax Division, reports prepared annually 1982 through 1989, Austin, TX

131. The 12- to 15-year criteria are not magic. The rationale is that when present value is calculated at an appropriate discount rate, the value of the annual income after 12-15 years becomes diminimus and adds little if anything to the property value.

132. In strong markets, particularly when royalty purchase companies or "trusts" are involved, the accepted period may increase to 5-7 years or more.

133. The book is derived from the transcription of a series of undergraduate Lectures on Physics during the 1961-63 period while Professor Feynman was teaching at California Institute of Technology. Professor Feynman presents six basic concepts of physics – atoms, basic physics, the relationship of physics to other topics, energy, gravitation, and quantum force – that are necessary to an understanding of theoretical and experimental physics. All in 136 pages. The book is both interesting and informative and would have been helpful when I was struggling through Marx Brook's Mechanics (Physics, not Auto Shop) class. There is a follow-on book titled "Six Not-So-Easy Pieces" that is more complex and something of a challenge. After some thought, it was decided that borrowing from the first volume was sufficient and would provide a similar collection of primary points related to the evaluation of oil and gas properties and projects.

134. xxx

135. *Manual for Discounting Oil and Gas Income,* Texas Comptroller of Public Accounts, June 2021, Austin, TX. Available from the website at comptroller.texas.gov/taxinfo/proptax

136. Assessors' Handbook 566-Assessment of Petroleum Properties, California State Board of Equalization, Sacramento, August, 1996. Available from the CSBE website at boe.ca.gov/proptax/ahcont.htm

137. *2021 School District Property Value Study, Discount Rate Range for Oil and Gas Properties,* Texas Comptroller of Public Accounts, Austin, TX, November, 2021

138. Weston, J. Fred and Copeland, Thomas E., Managerial Finance, 8th Edition, Dryden Press, 1986, p. 603

139. Mullens, David W., Jr.," Does *the Capital Asset Pricing Model Work?*" p. 283 in Financial Management - Harvard Business Review, John Wiley & Sons, Inc. 1983

140. Ibid, p. 283

141. As an example, *Value Line Investment Services* makes quarterly projections of earnings, etc., for a number of companies, but *Value Line* only covers a limited group of large companies.

142. *2021 School District Property Value Study, Discount Rate Range for Oil and Gas Properties,* Texas Comptroller of Public Accounts, Property Tax Assistance Division, Austin, TX, November, 2021

143. Prior to 2013 PTD calculated WACC for integrated and non-integrated companies separately.

144. "*WACC for Pure-Play Oil and Gas Extraction and Refining Entities,*" Ibbotson, Roger G. and Kaplan, Paul D., Ibbotson Associates, 1997, Chicago, IL

145. Ibid

146. Decision Analysis for Petroleum Exploration; Newendorp, Paul D. and Schuyler, John R., 2nd Edition, 2000 Planning Press, Aurora, Colorado

147. Decisions Under Uncertainty: Drilling Decisions by Oil and Gas Operators; Grayson, C. Jackson, Jr.;1960, Harvard University

148. Statistical Decision and Related Techniques in Oil and Gas Exploration; Kaufman, Gordon M., Prentice- Hall, Inc., 1963

149. An Introduction to Exploration Economics; Megill, R.E., Petroleum Publishing Company, Tulsa, OK. 1971

150. Risk Analysis and Management of Petroleum Exploration Ventures; Rose, Peter R.; AAPG, 2001

151. Risk, Uncertainty, and Investment Decision-Making in the Upstream Oil and Gas Industry; Macmillan, Fiona; University of Aberdeen, October, 2000

# INDEX

# G



# H



# I



# L



# M



# N



# O



# P

# R



# S



# T



# U



# V



# W